In-Process
Measurement and Control

MANUFACTURING ENGINEERING
AND MATERIALS PROCESSING

A Series of Reference Books and Textbooks

SERIES EDITORS

Geoffrey Boothroyd

*Chairman, Department of Industrial
and Manufacturing Engineering
University of Rhode Island
Kingston, Rhode Island*

George E. Dieter

*Dean, College of Engineering
University of Maryland
College Park, Maryland*

OTHER VOLUMES IN PREPARATION

In-Process
Measurement and Control

Edited by

STEPHAN D. MURPHY

Textron, Inc.
Danville, Pennsylvania

MARCEL DEKKER, INC.   New York and Basel

Library of Congress Cataloging-in-Publication Data

In-process measurement and control / edited by Stephan D. Murphy.
      p.   cm. --(Manufacturing engineering and materials
  processing ; 32)
    Includes bibliographical references.
    ISBN 0-8247-8130-9 (alk. paper)
    1. Process control. 2. Detectors. 3. Engineering
  inspection. 4. Quality control. I. Murphy, Stephan D. [date].
  II. Series.
  TX156.815   1990
  670.42--dc20   90-32072
                                           CIP

This book is printed on acid-free paper.

MARCEL DEKKER, INC.
270 Madison Avenue, New York, New York 10016

Current printing (last digit):
10 9 8 7 6 5 4 3 2 1

PRINTED IN THE UNITED STATES OF AMERICA

To my lovely wife

Nancy

and my beautiful children

Colleen and Brian

Preface

Inspection, including dimensional inspection, has commonly been an activity performed after, rather than during, a manufacturing step or process. In many instances, several steps may have been performed before a part is measured. If the part is found to deviate from the manufacturing blueprint tolerances, it must either be rejected at a point where considerable value has been added or be reworked. In either case, the manufactured cost has been boosted.

In view of the Japanese threat to American manufacturing, as well as internal competitive pressures, the American industrial community is turning to means of reducing manufacturing costs while improving quality. In addition to more standard approaches, such as increased automation and the use of robots, other more subtle approaches that work hand in hand with automation are being implemented. One such approach is in-process measurement.

In-process measurement is significant in that it ultimately allows a manufacturer to achieve a goal of zero scrap, since deviations in the manufacturing process measured by sensors can be used in a corrective manner to control the process before tolerances are exceeded. Advances in sensor technology and digital computers and controllers are permitting a dramatic increase in the application of in-process measurement and control.

Initially, digital computers were used for industrial control in the continuous process industries (e.g., petrochemicals). Although critics once believed that computers could not be used cost effectively by batch industries, they were ultimately proved wrong, as evidenced by the computers, programmable controllers, numerical controllers, and robots in use by these industries today. In the same way that temperature, flow, and pressure sensors are used in conjunction with digital computers for closed-loop control by the continuous process industries, dimensional sensors can be utilized within an in-process control loop for batch manufacturing processes.

This text attempts to encompass in-process measurement and control holistically as opposed to dealing with the bits and pieces, since the real power behind the concept comes from closing the control loop around the machine tool. It discusses not only various types of sensors but also strategies for using the data derived from the sensors in a closed-loop feedback arrangement. Also presented are discussions concerning the improvement in quality and reduction in cost achievable by in-process measurement and control through the use of numerical and programmable controllers. The text centers on the use of dimensional sensors; however, a logical extension of the concept reveals how nondestructive evaluation sensors could be used in a similar fashion to ensure material integrity.

Some of the material in this text has been excerpted from technical papers, while some reflects manufacturers' information concerning devices usable in an in-process control loop. The initial chapter deals with the title subject, and subsequent chapters discuss specific sensors, both contacting and noncontacting types. Later chapters discuss how data from the sensors can be used to close the loop around the process, and the last chapter discusses how in-process measurement and control can improve quality while reducing manufacturing costs.

This text is directed toward manufacturing and industrial engineers, technologists, and managers who are attempting to increase the level of automation in their factories while seeking ways to reduce manufacturing- and quality-related costs without compromising product quality. The concept of in-process control is especially important for those moving toward fully automated factories, since the on-line measurement capability also provides the process monitoring function previously relegated to the human operator.

ACKNOWLEDGMENTS

I wish to acknowledge the United States Air Force, who funded the Manufacturing Technology for Advanced Metal Removal Initiatives program that supported the work of Bob Thompson on air gauging and the work of Bill McKnight on the Tool Touch Auditron. Dr. Donald Flom of the General Electric Company was the program manager.

Steve Sahajdak's work on in-process gauging for a numerically controlled turning application was sponsored by the National Science Foundation. Professor Harry Mergler of Case Western Reserve University was the principal investigator on the project.

I would also like to thank all the contributing authors who made this book possible. Nowadays, in our busy, highly competitive business environment, it seems that there is barely enough time to do your job, let alone perform some extracurricular activities. I appreciate the effort put forth by all contributors.

Lastly, I wish to express thanks to my wife, Nancy, who assisted me in typing the manuscript.

Stephan D. Murphy

Contents

Contributors

FRANK C. DEMAREST   Zygo Corporation, Middlefield, Connecticut

THOMAS H. GEORGE   TRW Inc., Cleveland, Ohio

W. ANDREW HAGGERTY   Cincinnati Milacron, Cincinnati, Ohio

CURTIS D. KISSINGER   Mechanical Technology Incorporated, Latham, New York

ERIC KLINE   Cincinnati Milacron, Cincinnati, Ohio

WILLIAM S. McKNIGHT*   General Electric Company, Cincinnati, Ohio

---

Present affiliation: Belcan Engineering Services, Inc., Cincinnati, Ohio

STEPHAN D. MURPHY Textron, Inc., Danville, Pennsylvania

WALTER J. PASTORIUS Diffracto Ltd., Windsor, Ontario, Canada

STEVEN SAHAJDAK Consultant, Cleveland, Ohio

JAMES SOOBITSKY Zygo Corporation, Middlefield, Connecticut

ROBERT A. THOMPSON General Electric Company, Schenectady, New York

CARL A. ZANONI Zygo Corporation, Middlefield, Connecticut

In-Process
Measurement and Control

1
Concepts of In-Process Measurement and Control

STEPHAN D. MURPHY / Textron, Inc., Danville, Pennsylvania

> When you can measure what you are speaking about,
> and express it in numbers, you know something about
> it; but when you cannot measure it, when you cannot
> express it in numbers, your knowledge is of a meagre
> and unsatisfactory kind; it may be the beginning of
> knowledge, but you have scarcely, in your thoughts,
> advanced to the stage of science, whatever the matter
> might be.
>
> —Lord Kelvin

When William Thomson, Lord Kelvin, wrote these lines, he hardly suspected that the Japanese would one day surprise the American manufacturing community with low-cost, high-quality goods. He did, however, live during the beginning era of the industrial revolution, and, as the last line implies, these words can certainly apply to interchangeable manufacture. Measurement, particularly in-process measurement, is key for successful manufacturers. If manufacturers cannot measure their products' key features while the products are being made, they cannot begin to control their processes. This is fundamental to the implementation of in-process measurement and control and statistical process control (SPC).

The concept of in-process measurement and control has to do with measuring a process variable while that variable can still be influenced and applying a corrective feedback to the machine that affects the process so as to encompass the sources of error that normally occur during the process and thus eliminate error from the variable on the resultant workpiece. This text deals primarily with in-process measurement and control applied to batch machining operations, but the concept can be applied to other disciplines and is the basis for closed-loop control, whether it be a servomechanism loop, an engine speed governor, an analog amplifier feedback circuit, or a digital phase-locked loop.

As Schonberger has said (Ref. 1), one of the principles that the Japanese use to assure their high degree of quality is 100% inspection. In this way, the percentage of defective parts manufactured is measured in terms of parts per million instead of parts per hundred. There is a hitch to 100% inspection, however; simply stated, it is cost. Since the Japanese are in many cases the low-cost supplier, they have also addressed this issue. Schonberger states how devices called "bakayoke" are utilized to assure quality. These devices are the essence of in-process control, since they monitor the machine, measure the part, and warn the operator or supervisor of a malfunction or an imminent defective piece. As American manufacturers will learn, this approach of continuous in-process measurement coupled with tooling modification to permit fast changeovers will be the key to the implementation of just-in-time (JIT) procedures, which will allow the manufacturers to compete on par with the Far East.

Because of cost limitations associated with 100% inspection, the manufacturer must turn to a technological solution. A wide array of sensors are available which can be implemented to make dimensional measurements while the product is being made. In addition to these sensors, tremendous advances in the power of minicomputers, microcomputers, and computerized machine tool controllers have been achieved, along with an even more dramatic reduction in cost. These sensors, coupled with computerized control running appropriate software, can be utilized in a feedback arrangement to maintain the process within the allowable tolerance band. Consequently, extremely low defect rates can be achieved. Heretofore, the lack of sensors, the high cost of computers and associated controls, and the high cost of manual inspection precluded the use of in-process measurement

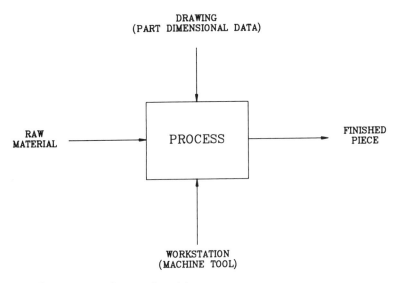

FIGURE 1.1 Basic metalworking process.

approaches. In addition, the in-process control solution involves a multidisciplinary approach requiring electronic engineers, computer scientists, control engineers, machinists, and computer numerical control (CNC) programmers. Many manufacturing shops do not have this type of talent under one roof. However, changes in American manufacturing are allowing many of the in-process measurement goals to be attained.

Figure 1.1 shows an IDEF (ICAM definition, where "ICAM" is integrated computer-aided manufacturing) model which represents the basic process and the basis for in-process control. In order to produce a part, raw material is required, a resource such as a machine tool is needed to effect the process, and something to describe quantitatively the amount of material to be removed—the part drawing or data from the drawing—is required. Normally, the engineering function of the manufacturing facility assures that the tooling design is done in such a way as to minimize changeover time for a new part and minimize operating loading variability. In some cases there may be an engineering trade-off between speed of changeover and the amount of variability in loading.

Once the part is loaded, the process can begin. If the parts are measured after the process, it can easily be seen that the dimensions will vary according to some statistical variation. This variation occurs because the process on the shop floor is not ideal. In addition to the variability in loading the part just mentioned, there is variance in the process due to deflection of the part, the cutter, the fixture, and the machine tool itself. In addition to the deflective variability, the different thermal coefficients of these members result in dimensional changes. There is wear of the cutting edge, and the material being cut may work-harden. Other problems may occur during the process, such as a catastrophic failure of the cutting tool or breakdown of a component in the machine tool. Moreover, many of the process parameters are interdependent and will change according to the spindle speeds, axes feed rates, amount of cutter engaged, type of material being cut, and depth of the cut. The result of all sources of variation may be a defective dimension and, ultimately, a defective part. For any particular process, the tighter the manufacturing tolerance, the more frequently dimensions will deviate, and any dimension that is not monitored and controlled can and will deviate. In many instances in manufacturing, the sources of variance are multiple and their interactions complex, to the extent that it is not practical to establish a deterministic process model for every operation that could encompass all sources of variability.

Therefore, it may be desirable to monitor and measure the dimension under question, in effect monitoring the performance of the machine while it is affecting this dimension. The measurement of the actual dimension can then be compared with the size mandated by the part blueprint and, finally, the machine can be adjusted accordingly to bring the dimension into the proper range.

Figure 1.2 expands on Figure 1.1 to illustrate this concept. As shown in the figure, the machine tool is fitted with a measurement sensor suited to the process and the type of dimension under measurement. This sensor provides a continuous measurement that can be in the form of an analog signal or a digital data word, which is compared with the required dimension derived from the part blueprint. The result of the comparison is a compensatory signal which is applied to the machine control so as to restore the dimension within its allowable range on either the part being machined or subsequent parts. The concept is directly analogous to a setpoint temperature controller used in a continuous process application where the setpoint is the desired or nominal temperature, the thermocouple is the measurement

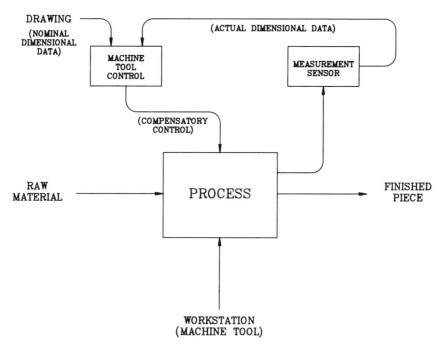

FIGURE 1.2   Concept of in-process measurement and control.

sensor, and the process is an oven.   The thermocouple measures
the temperature, which is compared with the setpoint.   An error
signal is determined, and the rate at which power is applied to
the oven is controlled accordingly to keep the temperature within
a desired band.   In the case of the machine tool, the setpoint
is the nominal blueprint dimension, which is being monitored by
a measurement sensor.   The difference between the two, the
error, is fed to the machine control, which activates the machine's actuators to restore the dimension within range.   Since
many of today's machines have computer numerical controls and
the requisite servomotor axis drives, the task of implementing
in-process control becomes one of selecting the appropriate measurement sensor and interfacing it to the machine tool control
with suitable hardware and software.

As will be discussed in later chapters, different types of
measurement sensors have different requirements for processing

the resultant measurement. Some have to do with the type of
signal that is acquired from the measurement transducer itself;
others have to do with the characteristics of the process. Since
every process and every sensor have some degree of random
variation associated with them, it is desirable to provide a degree of filtering or averaging of the measurements so as to remove these variances. In so doing, longer-term changes or
drifting of the process can be detected and corrected.

In-process control solutions need not be extremely sophisticated in order to be effective. The solution depends on the
type of process being controlled, the type of measurement being
made, and the rate of change of the process variables. Although
this text deals primarily with measurement and control applied
directly to a measurement on machine, the measurement sensor
need not be mounted on the machine. In instances where the
rate of change of the process variables is low enough—in other
words, the process is semistable—or where the tolerance band
is wide enough, the "in-process" measurement may actually be
made off the machine in a postprocess configuration as long as
the information is eventually sent back to the machine control
in the feedback arrangement discussed.

In many cases, however, the conditions are such that the
process could not tolerate a long phase delay without allowing
the dimension to exceed control limits. Under these conditions
the sensor may have to make measurements immediately at the
point where the dimension is being affected (e.g., at the tool-
workpiece interface). Moreover, neither the unit processing
the sensory data nor the machine actuators can introduce a significant delay without adversely affecting the degree of control.
Under these more critical circumstances, a full understanding of
the control loop is necessary in order to provide proper corrective feedback.

There are many possibilities for feedback control strategies
in an in-process control application. Most obvious is a direct,
inversely proportional correction. If the resultant error is
+0.002 in., for example, a correction of -0.002 in. can be applied to the control to compensate. As previously mentioned,
most processes have a degree of random variation which can be
described by a normal distribution. The allowable tolerance
band must be compared to the 6 sigma limits of this distribution
to establish whether or not the process is under control. If a

process is under control, one would normally base the feedback on an averaged value of measurements to avoid excessive compensation and possibly overcompensation of the process. This assumes, of course, that the variation is a significant percentage of the tolerance or control band. These averaged values can then be compared on a time scale to detect a trend or gradual shift in the output. A "rule of thumb" control strategy can then be set up according to these results. For example, a strategy may have the following rules: (a) if the result is within 50% of the tolerance band, do nothing; (b) if the result is within 50 to 100% of the tolerance band, do nothing unless three consecutive readings are strictly increasing or strictly decreasing (thus indicating a trend), in which case apply a correction equal to -1/2 (or some other multiplicative factor) of the error; and finally (c) if the result is outside the tolerance band, apply a correction equal to the inverse of the error.

For more sophisticated applications, the feedback could emulate a proportional integral derivative (PID) strategy similar to that used commonly in the continuous-process industries. Here, a proportional term applies an inverse correction with a gain factor associated with it, an integral term serves as a low-pass filter and removes quick statistical fluctuations, and a derivative term or high-pass filter can provide anticipatory or trend control. Lastly, if an analytical model is available which describes a complex process, corrective control can be applied as predicted by the model. For the vast majority of control applications on the manufacturing floor such as CNC milling or turning, however, the more simplistic approach as described in the previous paragraph will suffice.

There are many reasons why manufacturers should adopt in-process measurement and control philosophies in their plants. As we approach the fully automated factory, it is obvious why this type of approach will be necessary in the future. At present, there are many benefits that can be derived.

When a manual in-process gauging approach is adopted, it usually involves stopping the machinery in order to make a measurement. This stoppage results in unproductive time for the machine and the operator. It places a burden on the operator to keep adequate records and to do statistical computations in order to determine what amount of corrective feedback to apply. In many instances, the part must be removed from the machine tool and fixturing in order to make the measurement. When this happens, references may be lost due to refixturing problems in addition to the extra downtime to do the measurement operation. With the measurement sensor on the machine, readings are taken while machining, requiring no unproductive time. Data are collected automatically and can be stored as required while corrective computations are performed automatically. Measurements are not susceptible to human variability or operator bias. Relationships between dimensions can be held more closely with less refixturing.

In cases where each dimension is controlled, the manufacturer can approach a zero-defects condition. There will be fewer parts scrapped and a reduced amount of repair. Each of these reduces the cost of quality and will ultimately allow the manufacturer to reduce the labor content.

The in-process control approach also embraces the basis for an SPC program. Since measurements are being collected automatically, there is no labor cost associated with manual data collection. In many instances, a barrier to the implementation of SPC is the cost of the additional inspectors or additional unproductive time due to the operator's involvement. The initial reaction to the imposition of an SPC program is that the supplier is doing it on behalf the customer. In reality, the opposite is true; the customer is doing it for the benefit of the supplier, for the supplier who is required to implement SPC is forced to take measurements of the product, and by taking measurements, as Lord Kelvin implies, the supplier can start to understand the process. By understanding the process, the supplier can begin to control it, and only then can the supplier give the customer exactly what he or she wants.

At present, a change is taking place in the American manufacturing community. The issue of product quality and the role of the quality department are being addressed directly with a commitment for improvement from the upper-level managers down through the ranks. In many cases our historical view of quality was one in which quality control was done "after the fact;" that is, checks and measurements were not made during the machining process or in between operations but rather at the end of the manufacturing line, where it was too late to do anything about the immediate problems. In many ways this goes along with the philosophy of our society, which is wasteful and tolerant of waste. This goes against the grain of the Japanese, who are now our trading partners.

Our quality departments thus served the function of weeding out defective products, while the plant accumulated a large inventory to overcome the fallout. To make matters worse, we invested in labor to perform rework so as to recover the value added in our product when scrapping became too costly. All this despite the fact that the American government schooled defense contractors in the ways of SPC during World War II.

American products are thus competing against products from the Pacific Basin, whose manufacturers have not only lower labor costs, which is something we do not have much control over, but also lower operating costs, since smaller inventories are carried, fewer parts are scrapped, and labor is better utilized with fewer workers performing rework. At the same time, their customers love the products because they perform as advertised, they don't break down, and they last a long time—just what the customer wants and at a reasonable price. The solution to this problem is in how we view the quality function. Instead of the after-the-fact philosophy, we must adopt one in which quality is part of the manufacturing process, working from within to ensure that there is minimal deviated product.

Since we realize that a 100% manual inspection process would only drive up costs against competitors who can perform it with cheaper labor, we must consider automating the inspection process in the same way that we would automate a manufacturing process to reduce the direct labor content. To do this requires us to take advantage of existing measurement inspection sensors which can collect the data and apply them to the machining process along the lines of the feedback concept discussed in this chapter. For some applications, there may be an off-the-shelf solution. For example, Renishaw-type probes are adaptable to any machine tool with a tool changer and digitizing function built into the control. Although this type of arrangement does not allow continuous process monitoring, it does permit correction for subsequent operations and can be used to accommodate fixturing variance.

In summary, the need for in-process measurement and control has arisen owing to competitive pressure and market pull. The modern consumer or other end user of manufactured goods is now intolerant of inferior quality at a high price. Therefore,

manufacturers are obliged to ensure that the highest quality
standards are met while reducing price. What initially appears
to be a dilemma can be resolved by a change in the way in
which the quality function is viewed by company management.
In lieu of its historical role as a separate entity which segre-
gates nonconforming product at the end of the manufacturing
line, the quality function must be intertwined with the manufacturing process at every step along the way so as to control
every process variable within its dimensional allotment. In this
way, no product will deviate, and consequently complicated part
buffering schemes and costly excess inventory can be eliminated.
At the same time, there will be no need for the associated costs
of repair, rework, and scrap. With this in mine, it can easily
be seen how this philosophy of production can culminate in a
JIT implementation.

The other dilemmatic factor is the cost of implementing 100%
inspection. This must be addressed by a systematic application
of in-process measurement and control technology, since the
costs associated with manual inspection and data collection for
process control would be, in most cases, prohibitive. With the
increasing presence of computer numerical control and computerized process controllers on the manufacturing floor, the task of
implementing in-process measurement and control becomes one
of selecting the appropriate measurement sensor and designing
a suitable corrective feedback control strategy. There exist
measurement sensors that can be directly applied to the problem at hand, in addition to a bevy of sensors that can be utilized with a systems engineering approach to solve many production measurement problems.

REFERENCE

1.   Schonberger, R. J., "Japanese Manufacturing Techniques,"
     Free Press, New York, 1982.

2
In-Process Gauging Sensors

STEPHAN D. MURPHY / Textron, Inc., Danville, Pennsylvania

The purpose of the sensor is to obtain dimensional information from the workpiece. It is like a transducer in many instances because it converts one energy form to another. This other energy form is always an electrical signal, since we are considering sensors which provide an electrical signal to be used as feedback to the process or machine control. The sensor itself is key to the concept of closed-loop control since it must measure the desired parameter, not an ancillary one, and it must do so with the sensitivity needed for the degree of control or accuracy required.

Many of the sensors being considered here are not new; many have been utilized for years as stand-alone inspection sensors and have recently been modified for use as in-process sensors. The laser, which was at one time called an invention waiting for an application, has found widespread use in measurement sensors. With the reduction in price and improvement in semiconductor technology resulting in intelligent microprocessor-based signal processors, many sensors have found use as process sensors. One type of sensor, the video camera, could not be used without special high-speed data processing equipment.

In-process sensors for batch manufacturing can be broken down into two major categories: contact and noncontact. Noncontact sensors have come into greater favor because the lack of contact eliminates wear and deflection, which introduce inaccuracy into the measurement. This does not mean that contacting sensors cannot be used successfully. Contacting sensors such as linear variable differential transformers (LVDTs) have been used successfully in cylindrical grinding and turning operations (Refs. 1, 2). To some extent, the ability to use a contacting sensor is a function of the relative speed of the workpiece and the sensor. Where this speed is large, the potential for wear is extremely high, making contacting sensors less attractive. If an indirect measurement scheme (through the tool) is used, however, this contacting technique can be utilized despite large relative speeds.

Sensors can also be broken down into two secondary categories, direct and indirect, with reference to how the measurement is obtained. As mentioned in the preceding paragraph, measurements can be made indirectly. One example involves the use of a force sensor or an accelerometer attached to the cutting tool or cutting tool holder. In this case, to be used as a measurement sensor, the tool can be brought slowly up to the workpiece while the sensor signal is being monitored. At the point of contact with the workpiece, signals caused by minute deflections in the toolholder are amplified to be used as a strobe to record the position of the machine tool's carriage. An indirect measurement of the workpiece can thus be made by obtaining a reading through the machine tool's encoder at the moment the signal from the sensor is received. This method can also be used to derive additional information. Through the use of signature analysis—that is, examination of the signal in the frequency domain—information regarding the rate of material removal and surface finish can be obtained (Ref. 3).

An example of a sensor that provides a direct measurement is the LVDT. This sensor outputs a signal that is proportional to the distance moved; hence it gives a direct relative indication of position. When it is calibrated, an absolute measurement of the workpiece can be obtained independent of the machine tool's encoders.

A great variety of dimensional measurement sensors can be used successfully in an in-process application. The reader should be aware, however, that no one sensor is suited to all applications. Like part-holding fixturing and tooling, the in-process sensor is highly dependent on the part itself. This fact is most likely the reason for the limited use of in-process measurement sensing in industry today. To be able to apply an in-process sensor, the manufacturing, production, or tooling engineer is burdened with a multidisciplinary solution; the engineer must understand the electronics, signal processing, data processing, numerical analysis, and control theory in addition to the basic limitations of the sensor itself.

Available sensors have a variety of characteristics that must be understood before application. The major characteristics, physical size/standoff, resolution, accuracy, type of signal output, signal processing requirements, effective spot size, direct/ indirect, contact/noncontact, and limits of applicability are outlined herein.

The physical size of the sensor is important because it determines how close the sensor can be to the desired point of measurement. This is especially critical if the standoff distance, the distance between the workpiece and the sensor, is small compared with the sensor size. In a machining situation, the desired point of measurement is immediately behind the cutting tool. If the sensor makes measurements away from the cutter, this spatial separation results in a temporal delay before corrective control can be applied. Fortunately, certain conditions, such as cutter wear, that would cause a change in the measurement happen progressively, thus allowing a system to function with some degree of spatial separation. Modern intelligent motion control can accommodate this separation since workpiece dimensions are stored; hence the control can relate a measurement to its proper location. Other conditions, however, such as cutter breakage, require immediate action on the part of the control; hence it is advised that, in unattended machining operations, cutter force monitoring be utilized in addition to the dimensional sensors in an application where either the standoff distance is too small or the physical size of the sensor too large to permit measurement immediately behind the cutting edge.

Resolution and accuracy of the sensor are determined by the quality requirements of the part itself. Resolution is the smallest increment of distance that the sensor can resolve. In cases where the machine tool encoders are used to read location,

the resolution may be determined by the encoders themselves.
Repeatability, or ability to make the same measurement again,
is also a key characteristic of the sensor, as is accuracy, which
indicates how well the sensor can be integrated into the measurement and control system to permit calibration and, in turn, the
correct measurement can be made. One must keep in mind that
when there is some random variability at the measurement sensor which requires signal processing to maintain a certain accuracy, this processing can introduce time delays into the control
loop.

Different sensors have different types of electrical output.
Beyond the considerations of pure electrical compatibility, there
are basically two categories of output, analog and digital. Because of these two types of output and the requirement to assure
electrical compatibility, a unique interface must be specified along
with the sensor. Since we are considering including the sensor
in a feedback loop to a machine control, the sensor will be connected to a machine tool controller, which may be either a CNC
(computer numerical controller) or a PLC (programmable logic
controller). Most state-of-the-art CNCs and PLCs have a variety of optional interfaces (analog and digital), which allow the
sensor to be interfaced to the control without the need for custom interfacing. Moreover, these same controls have provisions
or optional modules which allow high-level language subroutines
to be written to handle the sensory data.

Under some circumstances, a special interface including a
significant signal processing is required for the sensor which
is beyond the capability of a machine controller. This is especially true for sensors such as the video camera. These specialized interfaces are normally designed around microcomputers,
personal computers, or minicomputers. As in the case of the
machine controllers, analog and digital interfaces are commercially available, making the chore of the electrical interface a
relatively easy one. Moreover, since the computer products are
a more mature technology, especially for applications where much
data processing and input/output (I/O) handling are required,
a wealth of software in the form of drivers is available to support the mechanics of the software interface, in addition to software to perform more sophisticated numerical analyses. In some
cases, specifically the video camera sensor, conventional computer interfaces may not be adequate. As discussed later, special-
purpose high-speed hardware processors are required to interface

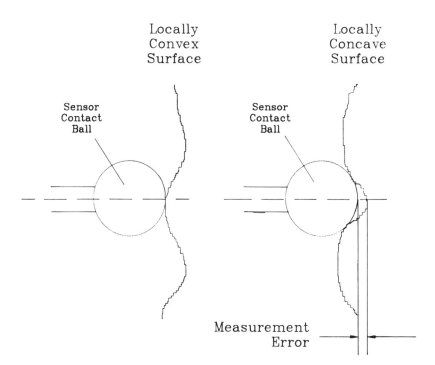

FIGURE 2.1   Effect of spot size on surface.

the sensor. To reiterate a previous comment, any data processing required will also introduce a time delay into the feedback
loop; therefore the amount of this delay must be reconciled with
the required response time of the loop.

The spot size or, more appropriately, effective spot size
refers to the active measurement area of the sensor on the workpiece, especially for noncontact sensors. The term is derived
from optical sensors that use a spot of light to perform the measurement. In terms of a contact sensor, it is the size of the
contacting member of the sensor. With respect to a micrometer,
for example, it is the diameter of the spindle and anvil. It is
important to appreciate the effect of the spot size, since it may
limit the accuracy of the measurement. This is especially true
where different geometries must be accommodated.

As seen in Figure 2.1, the effect of a ball contact can be
seen on convex and concave surfaces where the radius of the

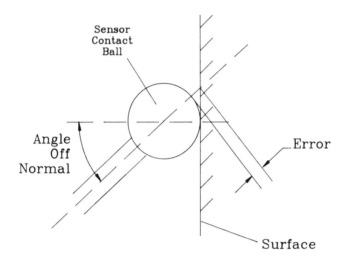

FIGURE 2.2   Effect of off-normal measurement.

ball contact is greater than the radius of curvature of the surface at the point of measurement.   Where conditions like this
occur, either the error must be tolerated or the spot size of
the sensor adjusted in order to reduce the error to a tolerable
level.   If the geometry of the part is known a priori, this resultant error of measurement can be computed and the value
adjusted accordingly.

     Another effect of the spot size has to do with the angle of
the sensor with respect to the surface normal at the point of
measurement.   This effect is shown in Figure 2.2.   Since the
sensor normally measures along its major axis, it has no sensitivity to displacement at angles other than zero degrees to this
axis.   Therefore, when this axis is not along a surface normal
of the object being measured, the measurement will be resolved
in other directions, resulting in another error.   Assuming no
other geometric considerations, this error can be corrected by
knowing the geometry of the sensor.   This correction is common
to all coordinate measuring machines using ball end contacting
probes.

     As discussed earlier, sensors can be broken down into four
divisions: contact, noncontact, direct, and indirect.   No one
sensor can be described as being the best overall, independent

of application, since in-process sensing is heavily application
dependent. Noncontact sensors have measurement accuracies
that meet or exceed those of contact sensors, yet in some applications the contact sensor may be more rugged and less affected
by severe environmental conditions present in machining. In
cases where wear is a consideration, the noncontact sensor is
an obvious choice. Moreover, the noncontact sensor cannot
mark or mar the workpiece, and in a sensitive measurement, it
does not deflect the workpiece. Indirect sensing is indicated
in applications where the measurement must be made immediately
at the tool. Direct sensors may have too long a standoff distance or be physically too bulky to make a measurement at the
tool. However, an indirect technique through the tool makes a
measurement through the machine tool's axes and is therefore
dependent on the tool geometry at a time when it is subject to
wear.

The balance of this chapter discusses a variety of sensors,
both contact and noncontact, which are used or can be used as
in-process measurement sensors. The sensing techniques to be
discussed are the LVDT, capacitive, eddy current, ultrasound,
air, laser, fiber optics, and video camera. These techniques
involve strictly direct measurement. The reader can see from
the list that it is weighted heavily toward the noncontact sensors. The noncontact sensor will undoubtedly gain in promin-
ence as the number of unattended machining operations increases
and quality standards become more stringent, requiring each machining operation to monitor carefully the part dimensions.

LINEAR VARIABLE DIFFERENTIAL TRANSFORMER

Perhaps the most mature technology for measurement sensors is
the linear variable differential transformer. The LVDT is normally configured as a contacting sensor, although several manufacturers have used the LVDT in conjunction with an air gauge
as a noncontact sensor. In this configuration, the air gauge
maintains a constant standoff distance from the workpiece while
the LVDT provides the direct measurement, thus taking advantage of the noncontact nature of the air gauge while using the
longer range capability of the LVDT. A schematic representation of the LVDT is shown in Figure 2.3.

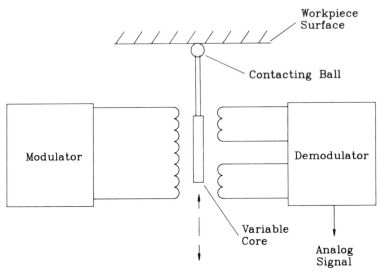

FIGURE 2.3 LVDT.

The LVDT, as its name implies, is a transformer with a variable core. It is wound with a primary coil and two secondary coils. The primary is wired to a modulator to provide an alternating-current source for the device. The secondary coils are connected to a demodulator, which measures the coils' output differentially in electrical opposition so that when the variable coil is centered within the transformer body, the electrical output is zero.

The output from the LVDT is typically a bipolar, linear analog signal. The interfacing circuitry required is an analog-to-digital (A/D) converter. Because the output is linear, no additional data processing is required, except in very precise applications, where small linear corrections can be performed using table lookup techniques to correct for nonlinearities in the demodulator, transformer windings, and coil placement. The digital output can subsequently be wired to a control via parallel or serial ports.

The LVDT has a maximum range of 3 to 5 in. Since its output is analog, its resolution is infinite, being limited only by the resolution of the A/D converter that follows it. Its accuracy is in the range of 0.001 to 0.000002 in., enabling its use in the most accurate applications. Since it is a contacting device, it has no standoff distance. Its effective spot size is determined by the contacting probe attached to the variable core. Small, hardened balls can be used with the sensor to minimize contacting errors. Because these balls can mar or score the workpiece when relative velocities are high, LVDTs have been fitted with rolling members which make them more applicable as in-process sensors. This sensor is a rugged, accurate, low-cost device whose only drawback is its contacting nature.

CAPACITIVE

The principle of this sensor is the variable capacitor. As shown in Figure 2.4, the sensor consists of the probe, one plate of the capacitor, the workpiece forming the second plate of the capacitor, and the associated processing electronics. Since the workpiece forms the second plate of the capacitor, the probe can be used with conductive workpieces. The sensor is noncontacting and provides a direct measurement.

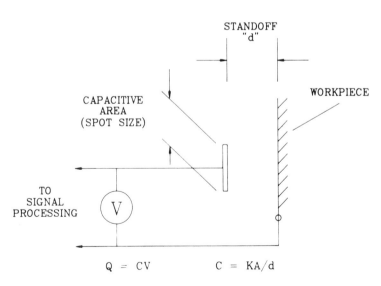

$$Q = CV \qquad C = KA/d$$

FIGURE 2.4 Capacitive sensor.

When a voltage is applied to the capacitor, an electric field is established between the plates of the capacitor. As the standoff distance d is varied, a proportional change in the value of the capacitance results according to the equation shown in the figure, assuming that the dielectric coefficient k and the area a of the capacitor's plate are constant. When the charge Q on the capacitor is held constant—no charge added or lost—the voltage V changes in proportion to the change in standoff distance.

The relationships shown apply to the theoretical limit where the area of the capacitor's plates is very large with respect to the standoff distance. In practice, this condition rarely occurs, since the plates would most likely be larger than the workpiece and certainly larger than a local geometric feature of the workpiece. The result of a measurement made under these conditions would contain a large error due to spot size considerations. In reality, fringe electrical fields at the edge of the capacitor can dominate the operation in cases where the capacitor plate is made small (in essence the spot size is small) to minimize geometric errors.

To make the spot size practical for in-process use, makers of these sensors have designed them with a type of guard ring around the periphery of the capacitor plate. The purpose of the guard ring is to minimize the effects of the fringe fields at the edges, thus making the capacitor behave as if it had infinitely large plates. Commercially available sensors have typical spot sizes in the range of 0.125 to 1 in. with measurement ranges of 0.010 to 0.5 in.

Because the output from the sensor is analog, the resolution is limited only by the A/D converter to which it is connected. Accuracy for this type of sensor is good; typically, accuracies of 0.0001 in. can be obtained. Use of this sensor is highly application dependent. This dependence is primarily related to the part geometry, which determines the maximum spot size for the application. Where small spot sizes are required, the measurement range, or standoff distance, is severely limited. In addition, a small spot size means a small capacitance. Hence, very sensitive amplifiers are required to detect the small changes in capacitance. Additional electrical noise resulting from the small signals translates into a larger filtering requirement.

In some applications, especially those where the geometry
is fixed, a larger spot size can be utilized to obtain a larger
standoff distance. Resultant geometric errors can be determined
and calculated out of the measurement. A major limitation of
this type of sensor is the fact that measurement range must be
sacrificed for spot size.

The measured output from this noncontacting capacitive sensor is a function of the standoff distance to the workpiece. This
distance may not be constant when the sensor is not normal to
the workpiece or when the surface under the active area of the
probe is not flat. As shown in Figures 2.1 and 2.2, an error
can be introduced due to lack of normality or surface curvature.
The capacitive sensor does not behave, however, in the same
manner as a contact probe. The voltage measured is based on
the average standoff distance, that is, the distance value obtained when integrated over the effective surface of the sensor.

When the sensor is used off normal against a flat surface,
the average distance is simply that measured at half the width
of the effective size of the sensor. Therefore, the sensor will
operate as if it were a point contact providing accurate readings
off normal. However, if the radius of curvature is small or if
local surface disturbances are encountered which must be measured, the sensor will also attempt to average these out; consequently, the resultant errors must be accounted for when they
are significant.

The capacitive sensor has moderate cost and good accuracy.
Its major limitation is standoff distance. Because of the standoff distance, the size of the sensor probe is sometimes compromised for this parameter or the sensor is used in applications
where the surface is flat. This compromise is accommodated in
applications where the surface is not flat by assuring a constant
geometry so that the errors can be subtracted. In a machining
situation, the sensor cannot tolerate grinding swarf or machining chips. In addition, water-based cutting fluids, since they
are conductive, cannot be tolerated. Although they are not
conductive, oil-based cutting fluids have a different dielectric
coefficient than air. This difference causes a change in capacitance indistinguishable from distance changes.

Primarily because of its intolerance to adverse conditions
during machining, the capacitive sensor is not suitable for in-
process measurements at or near the tool. It can, however, be

used in applications where the measurement is made postprocess, assuming that the resultant phase delay is not large with respect to the rate of change of the process at the machine tool.

EDDY CURRENT

The eddy current sensor is similar in concept and performance to the capacitive sensor. It does have some differences, though, that may make it more suitable in some applications. The major difference between the two sensors is the principle of operation.

The eddy current sensor utilizes an electromagnetic field as opposed to the capacitive's electric field. It is the electromagnetic field that makes this noncontacting sensor much less sensitive to the effect of cutting fluids. The sensing probe consists of a coil which is excited by a high-frequency alternating-current source. The resultant alternating magnetic field emanating from the coil generates eddy currents in the near surface of the material being inspected. These currents, in turn, create their own magnetic field, which couples to the coil, superimposing a current on the driving current. Demodulating circuitry detects this current signal, which can be calibrated to derive a distance measurement. As with the electric field, distance information is possible since the field strengths are a function of the distance from the sensor to the target.

Processing required for the analog output of this type of sensor is identical to that for the capacitive sensor. Resolution is limited by the A/D converter. Spot size, range, and stand-off distances are comparable to those available in a capacitive device. A conductive workpiece is required to support the induced currents.

Concerning spot size, the eddy current probe has limitations in terms of how small a coil, with enough turns to generate a sufficient magnetic field, can be wound. Since fringe fields have an effect on the output, the eddy current probe is also limited to constant-geometry situations where these errors can be determined and calibrated out. Since this device depends on eddy currents, which are a near-surface phenomenon, there are other variables peculiar to this sensor that must be understood.

The strength of the induced magnetic field is a function of the condition of the material. Because different materials have different resistivities, the sensor must be calibrated for a specific material. In addition, other material characteristics such as porosity and density affect the output. Near-surface conditions including defects such as cracks or inclusions alter the output and may be indistinguishable from changes in the distance being measured. Indeed, the primary use of this type of sensor is for surface and near-surface material defects, to which it is very sensitive.

This sensor, like the capacitive sensor, has relatively low cost and good accuracy, although it is highly sensitive to workpiece geometry and requires a means of calibrating out this variable. Similarly, its high sensitivity and accuracy must be traded off with standoff distance and limited range of operation. Different from the capacitive probe, it is not sensitive to the effects of cutting fluids and it requires a calibration for different materials.

ULTRASOUND

Sound waves can be utilized by ranging to obtain dimensional information. The configuration of the sensor involves a transmitter of sound energy and a receiver. In many cases the transmitter and receiver are in the same unit. Distance information is obtained by measuring the transit time required for the echo to return. Since the speed of sound in the medium is known, distance can be determined directly from time-of-flight information.

Ultrasonic energy, which is beyond the audible frequency range, is normally utilized since the wavelength of audible sound energy is relatively long compared with the resolution required for most dimensional measurements. For high-resolution applications, with measurement resolutions finer than 0.1 in., a liquid couplant (typically water) is required because the sound energy at the higher frequencies is highly attenuated in air. In an in-process measurement application, water- or oil-based cutting fluids can serve as a coupling medium. The workpiece under measurement need not be immersed as a continuous stream of liquid can adequately convey the sound.

The transmitter/receiver for an ultrasonic sensor is typically a piezoelectric crystal. The crystal itself is mechanically
damped in order to attenuate oscillation and avoid masking the
weak incoming echoes. The pulser is a fast, high-voltage
switch that drives the transducer with a short rise time pulse
that is converted into a mechanical pressure wave. The longitudinal sound wave is conveyed by the medium to the target,
where it is reflected back to the transducer. The same crystal
or an identical one converts the echo to an electrical impulse,
which is amplified by a tuned amplifier.

A timer is generally triggered by the pulse unit and dis-
abled by receipt of the incoming echo. The value held by the
timer is the time of flight of the sound wave, that is, the time
it takes the sound energy to travel to and from the target.
Since the speed of sound is constant for homogeneous materials
at a fixed temperature, the distance to the target can be determined by multiplying the speed by the time and dividing the
result by two to get a single path length. Digital counter/tim-
ers are presently utilized with ultrasonic sensors and provide
direct digital information for transmission over serial or parallel
data links.

The spot size of the noncontact ultrasonic sensor is determined by the size of the wave front of the sound energy. This
size is in turn governed by the configuration of the transducer,
which may have a lens incorporated to it to permit a focused
spot. This spot size may range from approximately 0.05 to 1.0
in. This type of sensor, like many of the other types of noncontacting transducers discussed here, tends to average the
information received from the target.

Different from other sensors, however, the returning sound
waves reflected from varying distances to the target integrated
over the spot size cause a dispersion of the time signal which
may result in trouble establishing an exact time reading, especially in cases where a constant thresholding circuit encounters
varying amplitude signals. This problem can be overcome, however, when the geometry of the workpiece is known or when the
sensor is kept normal to the workpiece. Under these circumstances, accuracies for this type of sensor of 0.0001 to 0.001
in. can be achieved.

Standoff distances in the range of a fraction of an inch to
tens of feet are possible with ultrasonic transducers. Again,

as previously inferred, with a long standoff distance and a lower-frequency sound wave, resolution must be compromised. In a machining application, accuracies mentioned in the previous paragraph may be realized with standoff distances in the range of a fraction of an inch to several inches. Like the eddy current probe, the ultrasonic sensor has the added benefit of being able to detect material defects. Unlike the eddy current probe, however, this sensor's sound energy can penetrate more deeply into the material and, by utilizing clever time discrimination of the resultant echoes, surface measurements may be distinguished from subsurface material defects, allowing the sensor to be used for both dimensional mensuration and material integrity of the workpiece.

AIR

The air gauging sensor is based on a fairly complex yet moderately priced technique which detects the difference between a reference pressure and a pressure through which air is bled between the sensor and the workpiece. The sensor is noncontacting and has the advantage that the air bleed through the measurement orifice keeps the part surface clean, making the sensor insensitive to the effects of coolant and thus an excellent candidate for in-process machining measurement.

A simple air sensor consists of a regulated source of air, an adjustable restriction, and a nozzle with an orifice, with an object, the workpiece, blocking the nozzle, the pressure across the restriction is zero. As the nozzle is moved away from the object, air flow commences and there is an increasing pressure drop until the point where the object presents no further restriction to the nozzle. This is the point where the nozzle is venting to atmospheric pressure and pressure drop is maximum. Over a certain distance when the nozzle is close to the object, there is a linear relationship between pressure and distance.

The most common type of air sensor is the differential air pressure gauge. The block diagram for the differential air sensor is shown in Figure 2.5. This configuration consists of a source of pressurized, filtered air and a two-legged air circuit. Each leg has an adjustable pressure regulator. The orifice leg has an adjustable restriction in series with the orifice that acts as an adjustment for the sensor's gain. In addition, there is an adjustable valve which acts as an offset. This valve bleeds

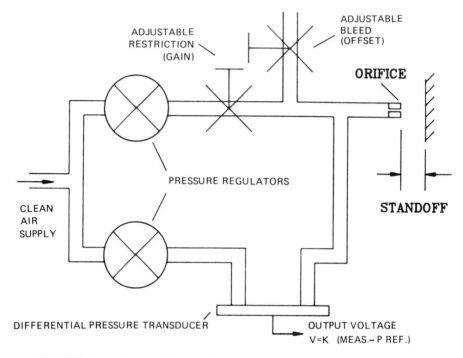

FIGURE 2.5 Schematic of air gauge.

air to the atmosphere, thus bypassing the orifice. A differential pressure is measured from a point behind the orifice to the output from the pressure regulator of the reference leg.

Within the standoff distance of the sensor there is a linear relationship between pressure and distance, enabling the sensor to be used as a distance-monitoring device. In this regime of operation, the area of the orifice is larger than the area surrounding the nozzle and between it and the workpiece. Due to this fact, the pressure drop is controlled by the area of air bleed between the nozzle and the workpiece. The differential pressure measurement, utilizing a reference pressure leg, makes the sensor insensitive to variance in the supply pressure.

For a typical air gauging sensor, the diameter of an orifice can be as large as 0.090 in. For this condition, a standoff range of 0.010 to 0.012 in. can be expected. The spot size is, for all practical purposes, the diameter of the orifice. As with many other noncontacting sensors, this sensor averages the distance across the effective area of the orifice. Hence, if off-normal geometry is encountered, it will measure true distance to the workpiece. However, when surface curvature or other surface features are encountered it will also tend to average these conditions and possibly introduce an error.

As previously stated, a linear response is measured across a differential pressure gauge. An analog voltage from the gauge would be converted to a digital word for processing by a controller, giving a theoretically infinite resolution. In reality, resolution and linearity to slightly better than 0.0001 in. can be achieved, making this sensor suitable for all but the most critical measurements.

The air sensor has many advantages in an in-process measurement scheme. It uses air, which is readily available at a machine tool site. When used in a differential mode, it is relatively insensitive to source pressure changes. Since the gauge operates on moderately high pressure, air flow through the orifice can remove cutting fluid films present on the part surface, allowing a true measurement of the part surface. This is one advantage over many other noncontact sensors. The working end of the sensor is very compact, consisting only of the nozzle and air supply tube. This compactness allows the sensor to be placed immediately behind a cutting tool, minimizing errors due to feedback delays (Ref. 4).

LASER

Several measurement methodologies, all noncontact, are possible using light sources. The utilization of laser light sources has made these methods increasingly sensitive due to the intensity, monochromaticity, and directionality of the laser beam. This is happening while the cost of laser light sources is decreasing. The two primary approaches to be discussed in this section are the laser triangulator and the shadow technique.

A traditional device improved by the use of a laser, the interferometer, can be used as an extremely accurate measurement sensor; however, it is not really suitable for on-machine use because of its high-cost requirement for stability and a specular reflecting surface. In an interferometer configuration (Michelson type), the laser's light beam is divided with a beam splitter into a measurement beam and a reference beam. The measurement beam is shone on the object. The reflected return beam is compared with the reference beam through a phase detector. Since in-phase light waves add constructively and light waves 180 degrees out of phase add destructively, the interferometer acts in principle like an optical encoder using the monochromatic light beam as an encoder scale. This enables it to determine relative motion to and from the object under measurement by counting light "fringes" caused by the alternating bands from the constructive and destructive interference. Multifrequency and Doppler techniques, which do not count such fringes, permit absolute measurements.

Using the interferometer approach, practical resolutions to 0.000001 in. are achievable with a standoff distance of a fraction of an inch and no theoretical upper limit. In reality, intensity of the laser, refraction of the light, and changes in the speed of light over long distances through the medium (the latter due to thermally induced changes in density in the primary medium, air) limit the standoff to several feet. For reasons discussed here, mainly lack of ruggedness and sensitivity, the laser interferometer has not found widespread use as an in-process measurement sensor. However, because of its extreme accuracy and resolution, the interferometer is being utilized to set up and calibrate machine tools and coordinate measuring machines. Improvements in this technique, such as those made by Hewlett-Packard to reduce the sensitivity to atmospheric changes, are making interferometer usage more widespread.

The two laser-based techniques that have been applied successfully as sensors operate on different principles, although both use the laser as a light source. The laser shadow gauge senses the blockage of a beam of scanned light, whereas the laser triangulator measures a spatial shift in reflected or scattered light to provide a measurement. Both types of sensors have a high degree of accuracy with reasonable standoff distances and small spot sizes.

The block diagram for the laser shadow gauge is shown in Figure 2.6. This type of system makes good use of the laser light source, which provides a monochromatic beam, obviating the need for achromatic optics, and has a small slowly divergent beam, allowing a small spot size over long standoff distances compared with other sensors. As shown in the figure, a laser

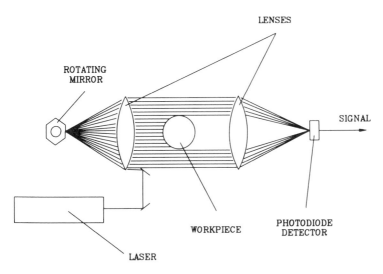

FIGURE 2.6 Laser shadow gauge.

is shone into a rotating mirror, which causes the light beam to
scan in a divergent manner. A lens is used to refract this divergent light into a plane of the highly parallel rays. The rays
can then intersect the workpiece under measurement within the
spread of the light plane, causing a shadow behind the object.
The edges of the shadow can be detected using a linear array
of photodiodes. Alternatively, as in the figure, which shows a
technique utilized by Zygo, an additional lens can be used to
collect the resultant image to be detected by a single photodiode.
With this approach, the size of the object is determined by the
time the photodetector "sees" the shadow and not by its spatial
size imaged on an array of photodiodes; consequently, the rotating mirror's speed must be tightly controlled to assure accuracy.
However, this is traded off against the time and complications of
additional electronic data processing imposed by the array.

  Because of the configuration of this type of sensor, only
features which are strictly convex can be measured. Other features may be shadowed by the part geometry itself, preventing
measurement. Accuracies down to 0.00001 in. are achievable
with this type of noninterferometric system. The field of view
of the sensor is limited to the size of the lenses utilized, which
for commercial units may range from 2 to 6 in. If larger objects

of variable size must be measured, the sensors may be adapted
to a positioning system. Where objects of fixed size larger
than the lenses are encountered, a configuration of multiple
sensors or single sensors with beamsplitters and mirrors may
be constructed.

The standoff distance of the shadow gauge is relatively
large compared with that of other noncontacting sensors. This
distance can be in the range of 10 in. for the accuracy previously stated. Although theoretically very large, the standoff
distance is limited by the accuracy of the lenses used to refract the light into parallel rays over the field of view. Moreover, at large distances conditions that affect the interferometer, such as atmospheric disturbances, will also affect this
type of system.

An advantage of this approach is that it is not affected by
surface condition or reflectivity as many light-based systems
are. In a machining application, it can be readily used in a
cylindrical grinding or turning operation (Ref. 4, 5). Although
it is bulkier than other sensors, through the use of mirrors it
is possible to direct the beam immediately behind the cutting
tool, minimizing delays in obtaining measurement data. Since
clear and translucent cutting fluids can refract the light at the
workpiece edge, an air stream or other means must be used to
remove cutting fluids for highly accurate measurements. These
types of systems are relatively insensitive to the presence of
cutting fluids and chips (unless they completely and continuously block the light path) since clever averaging and false-signal
rejection techniques have been implemented by the manufacturers.

The system can provide a measurement as fast as the mirror
can scan the beam. The effective rate may be several hundred
measurements a second. This rate, however, is reduced depending on the measurement environment and the accuracy required,
which determines the degree of signal rejection required. The
effective spot size is the width of the laser beam, which might
typically be 0.030 in. for an HeNe laser used in this application.
The sensor will tend to average a measurement, as opposed to a
micrometer used for an outer-diameter measurement, which would
read the peak. Also, imperfect alignment of the facets of each
reflecting surface in the rotating mirror provides a dither of the
beam, which tends to average the measurement.

Since the laser shadow gauge is normally supplied as a system and not only a sensor, the cost is generally higher than
that of most sensors. However, the units are easy to align,
provide a high degree of accuracy, are relatively impervious to
machining conditions (assuming that cutting fluids are not flood-
ing the measurement area), and provide a direct digital output
which can normally be interfaced to a computer or machine tool
control via a serial interface.

The laser triangulation technique, as the name implies, uses
a laser light source and the geometric principle of triangulation
to make a distance measurement. The principle of the sensor
is shown in Figure 2.7, where a reflecting approach is indicated
for clarity of explanation. The laser is commonly oriented normal to the surface and uses scattered light as opposed to direct
reflection as shown.

A laser light source and a linear photodiode detector array,
or a position-sensitive photodiode, are positioned away from the
workpiece with a known, fixed configuration. When light is reflected (and/or scattered) from the workpiece surface, it is

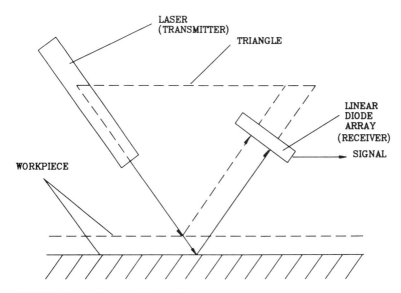

FIGURE 2.7 Laser triangulator.

directed onto the photodiode array. The photodiode array is
oriented so that when the workpiece surface moves from position A to position B, the reflected light follows a path along
the major axis of the array and the maximum light intensity can
be detected by another photodiode along the length of the array.

Since semiconductor processing techniques limit the size of
the photodiode, typically a positive lens is used in front of the
array to get a measurement resolution by magnifying the image
that is not limited by the diode spacing, which might be on the
order of 0.001 to 0.010 in. Actual resolutions of about 0.00005
in. are typical for this type of sensor. The spot size is the
size of the laser beam, and, as with other reflective-type light-
based sensors, the intensity of the signal is a function of the
surface characteristics of the workpiece material and the angle
of incidence of the light beam. Because of this, such sensors
must be able to accommodate a range of light intensities that
may approach 10000:1. This sensor, when used off normal,
will still provide an accurate measurement since it averages
across the beam area.

Because the angular configuration of the diode array is
known, as is the base of the triangle, the relative movement
from one position to another can be measured by locating the
position of peak light intensity along the length of the diode
array. This can be done numerically after scanning the array
and a distance from a reference can be output. Since diode
array lengths, as supplied, range from approximately 0.25 to
1 in. and a magnifying lens will reduce this effective length
by the magnification, the operational range of this type of sensor is small, on the order of 0.050 to 0.1 in. depending on the
resolution required and the diode array utilized. Therefore,
this sensor must normally be used with a positioning system
like other small-range and/or small-standoff noncontacting
sensors.
    The accuracies achievable with the laser triangulator range
from 0.00005 to 0.005 in. depending on the application. This
technique, like many others, can have its measurement accuracy
improved significantly through signal averaging and processing
techniques. Standoff distance ranges from 1 to 10 in. in typical applications with longer theoretical distances. Although the
early triangulators were fairly bulky units, primarily because
of the size of the gas laser used, the latest-generation sensors
are relatively compact, taking advantage of advances in solid-
state lasers.

Because this sensor is light based, the light path must be unobscured and the lenses kept clean when used in an machining application. The triangulator has been demonstrated in in-process control applications (Refs. 6, 7) and is supplied commercially by Diffracto, Selcom, and others. There is a trend among makers of coordinate measuring machines to equip the machines with this type of sensor, which will allow the machines to scan surfaces and measure points at a much higher rate than can be achieved with the existing contact Renishaw-type probes.

FIBER OPTIC

When one thinks about fiber optics the usual topic is communications. However, fiber optics have been utilized in many clever ways to produce measurement sensors for pressure, temperature, liquid level, and even gyroscopes (Refs. 8, 9, 10). One interesting application is the fiber optic lever, which can make an extremely sensitive measurement transducer. The effect was discovered by Kissinger and Smith (Ref. 11) and is described functionally in detail by Cook and Hamm (Ref. 12).

The fiber optic probe is a noninterferometric light-based sensor whose principle of operation is based on the fixed numerical aperture of the multimode glass fiber. Figure 2.8 shows the configuration of this noncontacting measurement transducer. The sensor consists of a bifurcated fiber bundle with one leg connected to a photodiode and the other to a light source. The light source can be white light or laser (Ref. 13). In view of the large range of light intensities reflected from machined surfaces and the reduced light levels when used off normal, the increased intensity and better coupling of the laser to the fiber bundle make it the preferred source.

At the point of merger of the two bundles the fibers from the source end and the detector end are randomly mixed, although other configurations with concentric and hemispherical arrangements will still function. Normally the sensing end is coupled to a noninverting optical system which displaces the focal point of the sensor, creating an erect virtual image on the object being measured, and, if desired, demagnifies the overall active bundle, in effect reducing the spot size.

The principle of operation is best explained without lenses as shown in Figure 2.9. The figure shows one transmitting

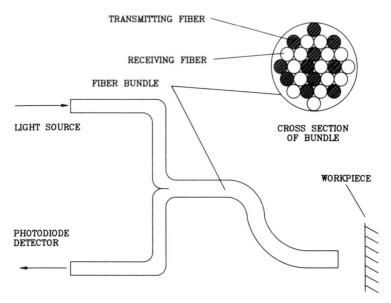

FIGURE 2.8   Fiber optic sensor.

fiber with two receiving fibers alongside it. With the probe
against the object, no light is received by the photodetector
coupled to the receiving fibers, and there is a minimum signal
at this point. As the probe is withdrawn from the object,
light is reflected and scattered from the surface and is directed
toward the receiving fibers. Because an optical fiber has an
aperture, that is, an angle beyond which the light rays are
neither accepted nor transmitted, there will be a preferred
standoff at which light coupling from the source fiber to the
receiving fibers reflected from the part will be maximal. At
this point the signal from the photodetector will be at its maximum value. As the sensor is retracted farther from the object,
the cone of light emanating from the fiber is coupled outside
the receiving fibers and consequently the light intensity at the
detector is reduced.

Between the minimum and maximum signal levels there is a
linear relationship between position and signal level due to the
geometry established by the aperture angle of the fiber. Because of this aperture angle, the light coupled into the receivers is levered; that is, the amount of light coupled into the

RECEIVING FIBERS   WORKPIECE
FACE OF FIBER
BUNDLE

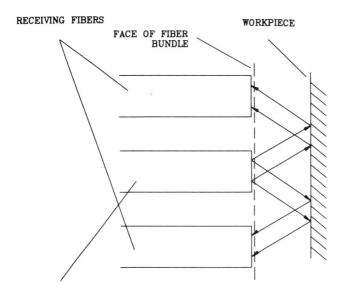

TRANSMITTING FIBER

FIGURE 2.9   Principle of operation of fiber optic sensor.

receivers is extremely sensitive to position. Consequently, the
configuration allows this sensor to be used as a very accurate
measurement sensor.

   More significantly, when coupled to an optical system as
previously stated and used with a larger random array of fibers
as illustrated in Figure 2.8, a virtual image of the end of the
fiber bundle can be projected onto the part with a typical spot
size of 0.020 in. When lenses are used, the signal level increases as the sensor is drawn away from the object until coupling of the virtual image of the transmitting fibers is optimized,
resulting in a local maximum. When the sensor is withdrawn
farther, the signal level drops sharply to a local minimum at
the focal point of the optics as the images of the transmitting
fibers are focused on themselves and minimal light is coupled
to the receivers. Continuing farther on the opposite side of
the focal point, another local maximum is obtained at a slightly
lower magnitude than the first due to the increased distance
from the part. The standoff distance is normally measured at
the focal point. With lenses, this standoff can easily be in the
range of 0.25 to 3 in.

Although this signal is linear in the region of the focal
point and can be used in this manner in a measurement sensor,
the light detected by the photodiode is a function of the light
source, which could well be time variant due to noise, power
supply modulation of the laser tube, or use of a white light
source. Consequently, the sensor is best used in conjunction
with a positioning system that locates the local minimum (at the
focal point of the optics), whose position is independent of
source light variation. When used in this manner, practical
accuracies of 0.00001 in. can be achieved with resolution dependent only on the positioning system used. Si ice the sensor
is noninterferometric, accuracies even better tha the one
stated could theoretically be achieved. As with .he other light-
based sensors, some degree of signal averaging is required
especially when high accuracy is expected; this accuracy must
be traded off against measurement speed.

In summary, this type of sensor can provide similar accuracy and performance to triangulators and interferometers without the sensitivity of the interferometer. Like the triangulator,
it requires a positioning system to give a workable measurement
range; therefore an intelligent processor is required to locate
the focal point and relate it to an encoded position so that a
distance measurement can be made. The response time of the
sensor, which may typically be 1/4 to 1 second per measurement, is limited by the response of the positioner. Since it is
a light-based sensor, the optics must be kept relatively clean
and unobscured from the workpiece being measured to permit
an accurate measurement. Also like the triangulator, which
depends on a reflection from the workpiece surface, this sensor must accommodate a wide range of light intensities due to
differences in angle of incidence and surface finish and color.
Normally, logarithmic compression of the light signal can handle the range of intensities encountered in a manufacturing
environment.

VIDEO CAMERA

Perhaps the newest, most flexible noncontacting sensor, and
the one with the most potential for the future, is the video
camera-based measurement sensor. These sensors are normally
categorized under the title of machine vision. The technology
is based on a video camera (vidicon) and a computer-based

image analyzer. The technology of the components is fairly mature, dating back to television and the digital computer. Ironically, machine vision has been used in an industrial setting only within the past five years. Its use is primarily due to the availability of stable solid-state vidicons, faster, less expensive minicomputers, and image processing algorithms. A significant advantage of this approach is that an entire scene can be captured in an instant, allowing a multitude of measurements to be made without moving the sensor. The main disadvantage stems from the volume of data that must be processed by the image analyzer, resulting in either a slow response or a requirement for an extremely fast processor. Although a camera can acquire only a two-dimensional scene, structured lighting approaches combined with triangulation can be implemented to allow the video camera-based sensor to collect measurements in three dimensions.

The component parts of a machine vision sensor are simply a vidicon, interface, and computer. In practice, a configuration as shown in Figure 2.10 would be implemented. The video camera views the workpiece under measurement through an optical system, normally a magnifying lens. The sensing element is a solid-state photodiode array, although, in principle, a conventional vidicon tube can be utilized. The solid-state sensor has many advantages, including lower cost, ruggedness, less sensitivity to blooming, wider dynamic light range, greater image stability, smaller size, less weight, and ease of interfacing. The spacing of the photodiodes is typically on the order of 0.001 to 0.005 in., which would limit the resolution of the sensor. For this reason, magnifying optics are usually used to achieve resolutions in the range of 0.0001 to 0.010 in., which would normally be required in a manufacturing environment.

The camera is connected to a frame grabber, which is a specialized, high-speed memory designed to collect and store single or multiple images, called frames, from the camera. For each frame the frame grabber stores, it must have a capacity of at least the same number of photodiodes as the camera has in its array. The individual diodes are called pixels, for picture elements. The frame grabber, in turn, must accommodate the same number of pixels. Since the photodiode itself merely provides a light level as an analog signal, this must be converted to a digital value for the memory and the digital computer. This conversion is normally done by means of an analog-

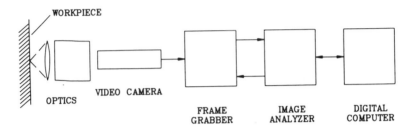

FIGURE 2.10   Block diagram of video camera-based measurement sensor.

to-digital converter. The A/D converter resides in either the camera or the frame grabber. The resolution of the converter determines the number of levels of light that can be resolved. This range of levels is normally termed the gray scale and the typical resolution is 8 bits.

The frame grabber is connected to an image analyzer, which is another piece of specialized circuitry designed to unburden the digital computer from handling all the data collected from the camera. Repetitive, high-speed processing of the data can be performed by the image analyzer breaking the picture down into features for further analysis, including measurement by a less costly computer of moderate speed. Operations that the image analyzer might handle include averaging, convolution, edge detection, thresholding, and windowing. Operations for image analysis originally developed at SRI (Ref. 14) and GM (Consight) spurred use of the video camera for inspection purposes.

To utilize the camera as a measurement sensor, features in the field of view of the camera must be located. This is normally done through edge detection. Edge detection, as the name implies, is used to locate the edge (light-to-dark transition) of an object so that it can be used as a measurement point itself or so that it may be used with other points to describe a feature such as a line, circle, arc, or ellipse. For objects that cannot be contained within one field of view, points obtained in one frame can be used to reference other views. Edge detection is a differentiation process (e.g., Sobel operator) usually done by examining a small region of pixels for their light-to-dark transition in the vicinity of the edge.

Since differentiation is inherently a noisy process, normally it is combined with some averaging, statistical, or convolution techniques to "clean up" the image. After edge detection, the resultant image can be subtracted from a threshold to establish a clean edge.

After points on the edges are found, they can be fit—for example, by using a least squares approach—into lines (features) that describe the object. A least squares approach has the advantage that it is an averaging process which aids in removing random variation that might affect feature location. The features can then be passed up to the computer, where measurements can be made by subtraction of absolute feature locations of preselected points. Where objects are too large for the field of view of the camera and lens combination, the computer can combine images into data that describe the entire object. Features of the object might also be directly compared with entities from a computer-aided design data base, for example.

Because of limitations in the size of photodiode arrays and a reduction in field of view due to the magnifying lens required to get proper resolution, in many cases a positioning system is also required for a video camera sensor. When a large object is measured, the camera (or object) is moved in the vicinity of the points to be measured. In some cases a third axis of motion may be required to focus the camera. Again, when magnifying optics are used, the depth of field of the system, like the field of view, is reduced. When few points must be measured, the image can be windowed to improve the response time of a measurement. Windowing reduces the area of interest in the scene to the area where the point to be measured is expected to be found, thus reducing processing time greatly.

Accuracies in the range of 0.0001 in. can be achieved with a video camera-based sensor. Standoff distances may range from 0.050 in. to several inches, with the former number typical of a microscope objective lens. Other optical solutions are available in the form of macrofocusing telescope lenses where standoff distances on the order of feet may be obtained with magnification. Limitations due to spot size associated with other sensors do not exist, since the camera can view a whole scene at once. As with other light-based sensors, limitations due to dirty lenses and objects blocking the lens such as coolant and chips apply to this camera-based sensor.

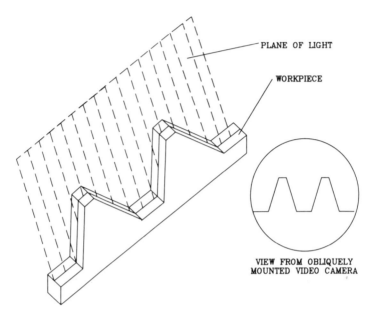

PLANE OF LIGHT

WORKPIECE

VIEW FROM OBLIQUELY
MOUNTED VIDEO CAMERA

FIGURE 2.11   Example of structured light illumination.

Proper illumination of the object is critical to successful application of a vision-type sensor. Improving the illumination can also reduce the data processing required by reducing the uncertainty of edge detection. The sensor can utilize both back-lit and front-lit options, depending on the application. It is readily apparent how a back-lit application of this type of sensor could automate an inspection operation by emulating an optical comparator (Ref. 15).

An extension of this sensor which permits determination of three-dimensional information involves the use of structured light scenes. Figure 2.11 shows how an object can be measured using structured light in a light sectioning example. The light source can be a scanned laser or a laser shone through a cylindrical lens to create a plane of light which intersects the object under measurement at a plane different from that of the camera. As seen in the figure, the camera sees an image of the intersection of the plane of light and the object. Once the image has been corrected for the geometric transformation caused by the off-angle perspective, the image processing techniques discussed earlier can be applied to extract dimensional information. Now, if the light source and camera are moved in a direction normal to the plane of light, another section of the object can be measured, providing information in the third dimension.

The video camera-based sensor can also be utilized in the same manner as the laser triangulator, the main difference being the use of the two-dimensional array as opposed to the linear array used in the triangulator. In this manner, three-dimensional information can also be obtained from a scene. With this approach one issue must be addressed when attempting to measure three-dimensional data that come from mapping a three-dimensional scene onto a two-dimensional plane. This issue does not cause confusion in the laser triangulator, since there is enough a priori knowledge about the scene.

In the general case, when mapping a three-dimensional scene onto a plane, a ray of light emanating from a point on a close object may be mapped onto the same pixel in the camera's photodiode array as a ray emanating from a point displaced from the first on a distant object. This confusion must be resolved in order to obtain a distance measurement. To do this, another type of structured light application, described by Altschuler et al. (Ref. 16) can be used to make measurements in three dimensions from an unknown scene. In this application a shuttered, structured lighting system has been developed to control the illumination of the object being measured, thereby permitting a one-for-one mapping from the three-dimensional scene to the two-dimensional image. To accelerate the data processing, a binary sort routine was utilized to analyze structured light patterns from the shutter.

As can be inferred from the previous discussions, the video camera-based measurement sensor is extremely flexible for in-process measurements. Its accuracy is well within the range for all but the most demanding applications and is certainly comparable to that of other contacting and noncontacting sensors. The camera can be configured with a variety of lenses and light sources to accommodate different field-of-view and standoff requirements. The primary disadvantages of this type of measurement sensor at this time are its cost, complexity, and response time for complicated scenes; however, this will change as the technology matures and less expensive high-speed computers become available along with specialized very high speed integrated circuit (VHSIC) chips to perform the image analysis functions. It should be noted that these sensors are finding their way onto the manufacturing floor, where they have provided viable cost-effective solutions to some tough in-process measurement problems.

In conclusion, a variety of sensors have been presented here which can handle the function of providing in-process measurement data. One statement can be made regarding the application of a sensor to an in-process measurement require-ment: no one sensor can be called on to solve all problems. It has been the goal of this chapter to present descriptions of various types of sensors in the hope that the reader can make a judicious selection for a particular measurement problem. The reader should also be aware of the trends toward noncon-tacting sensors that at this point can outperform conventional gauging sensors for reasons that have been discussed here.

REFERENCES

1. G. Branchini, "Gaging and the Closed Loop," Tooling and Production, August 1979.
2. G. Branchini, "Automated Taper Correction for Cylindri-cal Grinding," Tooling and Production, May 1981.
3. D. A. Dornfield, "Tool Wear Sensing via Acoustic Emission Analysis," U.S. Department of Commerce, Rotating Tool Wear Monitoring Apparatus, PB82-197708, April 1982.
4. D. G. Flom, "Manufacturing Initiative for Advanced Metal Removal (Task IV, In-Process Inspection, R. A. Thomp-son), F33615-80-C-5057, October 1984.
5. A. Novak, "Quality Control of Turned Workpieces by Means of Laser," 6th International Conference on Auto-mated Inspection and Product Control, Birmingham, England, April 1982.
6. H. W. Mergler, "In-Process Optical Gaging for Numerical Machine Control and Automated Processes," NSF Grant DAR 7821639, September 1979.
7. H. W. Mergler and S. Sahajdak, "In-Process Optical Gag-ing for Numerical Machine Control and Automatic Processes," Proceedings of the 3rd International Conference on Auto-mated Inspection and Product Control, 1978.
8. J. Hecht, "Fiber Optics Turns to Sensing," High Technol-ogy, July/August 1986.

9. A. R. Tebo, "Sensing with Optical Fibers: An Emerging Technology," Electro Optical Systems Design, February 1982.

10. M. Sharma and R. E. Brooks, "Fiber Optic Sensing in Cyogenic Environments," SPIE Vol. 224, Fiber Optics for Communication and Control, April 1980.

11. C. D. Kissinger and R. C. Smith, "Improved Non-Contact Fiber Optic/Lens Displacement Measuring System," Proceedings of Electro-optical Systems Design Conference, July 1978.

12. R. O. Cook and C. W. Hamm, "Fiber Optic Lever Displacement Transducer," Applied Optics, Vol. 18, October 1979.

13. Automated Non-Contact Airfoil Gage, Compressor Components, Textron, Danville, PA.

14. R. C. Bolles, J. H. Kremers, and R. A. Cain, "A Simple Sensor to Gather Three Dimensional Data," SRI, July 1981.

15. R. Cederberg and S. Svensson, "Optical Automatic Control for the Lathing of Parts," FDA Report C 30175-E, October 1979.

16. B. R. Altschuler, M. D. Altschuler, and J. Toboada, "A Laser Electro-optical System for 3-D Topographic Mensuration," Proceedings of SPIE Technical Symposium in Imaging Applications for Automated Industrial Inspection and Assembly, April 1979.

3
Laser Telemetric System

FRANK C. DEMAREST, JAMES SOOBITSKY, and
CARL A. ZANONI / Zygo Corporation, Middlefield, Connecticut

INTRODUCTION

Manufacturers of any product today must produce higher quality and lower cost to compete effectively in world trade. The improved quality and higher accuracy of components and raw materials needed to build a world-class product have moved the quality control inspections closer and closer to the manufacturing processes. This allows the dimensions of the part being manufactured to be monitored as close to its point of manufacture as possible so that the process corrections can be made as soon as an error is detected.

A large number of manufactured components and raw materials are cylindrical in shape and are manufactured in continuous processes. The materials of these components range from steel to rubber, plastic, and glass. In addition to the various types of materials, the motion and temperature of the components in process can vary over a large range. The preferred method for measuring soft, delicate, hot, or moving objects is with a noncontacting sensor. A few types of noncontacting sensors are capacitive gauges, eddy current gauges, air gauges, and optical gauges. Optical sensors have a number of advantages: They are truly noncontact, in that there are no forces on or connections to the part being measured. They can measure

parts of any material. They allow the distance from the sensor
to the object being measured to be large. Various techniques
are used in optical dimensional gauging. These include shadow
projection, diffraction phenomena, linear arrays, and scanning
laser beams. The scanning laser beam is the method the Zygo
Corporation chose for the Laser Telemetric System (LTS). This
chapter presents a design study of this laser-based measurement
system suitable for industrial in-process measurement.

HISTORICAL OVERVIEW

In 1972 two manufacturers communicated with Zygo looking for
a noncontact optical sensor to measure extruded and rolled material on line for in-process quality control. Their requirements
of the sensor were 2 in. (50 mm) measurement range, 0.0002 in.
(0.005 mm) accuracy, and a 4 in. (100 mm) measurement throat
which could tolerate a large motion of the part without loss of
accuracy. At that time the other manufacturers of scanning
laser beam sensors did not offer an instrument of these capabilities. Zygo felt that the development of such a sensor was within their capabilities and that a sufficient market existed for the
sensor to pursue its design. The first commercial LTS from
Zygo was introduced in 1975. The LTS had a 2 in. (50 mm)
measurement range with 10 in. (250 mm) depth, 0.0002 in.
(0.005 mm) accuracy, separable transmitter and receiver, and
no need for frequent recalibration. This system relied on the
optomechanical design to give the required system accuracy.
The exact edge of the part was sensed using a technique of
electronically produced derivatives of the light signal, for which
Zygo was issued a U.S. patent and which it still uses today.
With the advent of microprocessors the LTS evolved a new electronic design which allowed the use of different optical systems
and increased the accuracy of the instrument. Also, the new
software provided special process control functions and internal
computation capability making it possible for the LTS to supply
the user with corrected data directly. The current state-of-
the-art LTS provides the user with many features. The sensor
types range from 1 in. (25 mm) to 18 in. (460 mm) in measurement range with corresponding accuracies of 0.00003 in. (0.0008
mm) to 0.0005 in. (0.013 mm). The controller features standard
software for statistical process control, histogram of data trends,
multiple measurements, user-definable internal calculations,
transparent object measurement, and expressions which use data

from two or more sensors. Also, the controller can drive up
to four sensors internally with the option of monitoring other
sensor types such as temperature probes or linear encoders
through the computer interface.

DESIGN CONSIDERATIONS

In order to evaluate the relative advantages of a scanning laser
beam sensor and to understand its limitations, the concept, environmental requirements, and error sources related to this
method of measurement must be examined.

The concept of the LTS is very straightforward, as shown
in Figure 3.1. The basic function of the instrument is to provide high-speed, noncontact precision measurements of various
types in many different environments. The system shown in the
figure is the schematic representation of the optomechanical design used for the first LTS. The laser beam is directed through

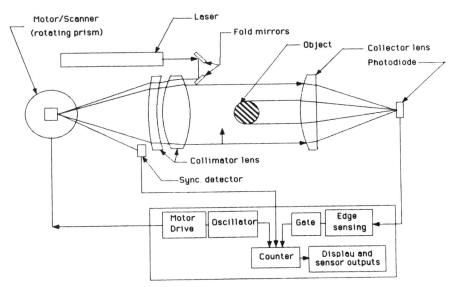

LASER SCANNER FOR DIMENSIONAL MEASUREMENT
(simplified schematic models 110,120)

FIGURE 3.1 Laser scanner for dimensional measurement.

the collimating lens and is reflected off a rotating mirror, known
as the scanner, to produce an angularly scanned laser beam.
The surface of the scanner is located at the focus of the collimating lens which converts the angularly scanned laser beam into
a parallel scanned laser beam. This parallel scanned beam is
partially blocked by the object under measurement, which cre-
ates a shadow of the object. The shadow is actually a time-
varying loss of light. What seems to the eye to be a static
shadow projection is in reality a time-dependent on-off-on again
signal when the shadow is viewed as a function of time. The
parallel scanned laser beam is focused by a collector lens onto
a photodetector which converts the light signal to an electrical
signal. This signal is amplified and differentiated twice to produce the electrical representation of the exact center of the
beam as it passes over each edge of the part. The time between these edge signals is measured and corrected, and the
dimensional information concerning the object is obtained. The
dimensional output is calculated in only a fraction of a second.

   The internal components of an LTS are normally grouped
into three distinct subsystems: the transmitter, the receiver,
and the controller, as shown in the diagram. The transmitter
houses the laser, the scanner/motor, the optical bar, the collimating optics, the sync. detector or autocalibration mask
(auto-cal), and the beam-shaping optics. The receiver houses
the collector optics, the photodiode, and the preamplifier with
digitizing electronics. The controller contains the remaining
electronics for the measurement calculations, such as the reference clock, the correction table, the operating software, and
the power supply for the motor.

   The LTS is used in a variety of environments, ranging
from an inspection room to the rolling line of a steel mill. A
requirement for the instrument was to have a separable transmitter and receiver and to allow the object being measured to
be placed anywhere in the measurement space with no degradation of performance. The sensor must be insensitive to mounting orientation and must provide for a simple mounting technique.
Although the concept of the LTS is quite simple, these environmental and mounting requirements have made implementing it
into an industrial product (i.e., one that works on the shop
floor) with the required performance a significant technical
challenge. The following brief look at the advantages of the
LTS shows the usefulness of this type of instrument:

1. Noncontact measurement allows the LTS to measure moving, delicate, soft, hot, radioactive, and, with special software, transparent objects.
2. The position of the object being measured can be anywhere in the measurement volume of the instrument. This means that the object can move as its dimension is being measured and the distance between the sensor and the object can be anywhere from several inches to several feet.
3. More than one object can be measured simultaneously.

SOURCES OF ERROR

This section is dedicated to the description of the error sources involved with this type of laser-based measurement. As with any type of metrology, error sources must be identified and eliminated or corrected in order to obtain an accurate, repeatable measurement. As noted previously, the handling of the various sources of error was a significant technical challenge that resulted in the development of a successful industrial product.

Measurement errors can be introduced either by the instrument used to make the measurement (internal sources) or by the geometric or environmental factors related to the particular measurement situation (external sources). These errors, whether internal or external, are either systematic or random in nature. Systematic errors cannot be reduced by averaging the data, whereas random errors can be reduced. The ultimate performance of the system is, of course, based on the composite effect of both types of error sources.

The primary external error sources are listed below along with their type:

Error source	Type
Alignment of object measured	Systematic
Motion of object measured	Systematic and random
Atmospheric effects	Random
Dirt	
In measurement region	Random
On object	Systematic

Error source	Type
Temperature	Random
Surface finish of object	Systematic
Edge-sensing errors	Systematic
Stray light	Systematic and random
Process effects	Systematic and random

Alignment of the object being measured affects the measurement. If the object is tilted with respect to the scan line of the measurement beam, a geometrically induced error as depicted in Figure 3.2 appears. The geometric error is given by

$$e = d\left[\left(\frac{1}{\cos a}\right) - 1\right] \qquad (3.1)$$

where d is the dimension of interest and a is the tilt angle. Equation 3.1 shows that for a constant error the allowable tilt decreases as the dimension measured increases.

Since the dimension of the object is directly related to the time it takes to scan the object, an error is introduced for a moving object. If the object is moving along the axis on which the beam is scanning, an error will be induced which is given by

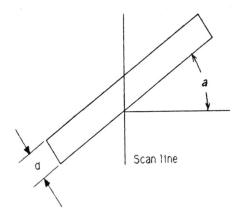

FIGURE 3.2   Error induced by alignment.

$$e_m = d \left| \frac{V_{tp}}{V_b} \right|$$
(3.2)

where d is the dimension being measured, V_{tp} is the transverse
velocity of the object being measured, and V_b is the velocity of
the measurement beam. It is important to realize that only the
velocity component of the part parallel to the measurement beam
motion is a source of error. If the transverse part motion is
unidirectional, this error is systematic. If the part motion is
vibratory—that is, it moves both along and opposite to the
velocity of the measurement beam—this is random in nature and
can consequently be reduced substantially by averaging.

Atmospheric effects (thermal and turbulence) can affect the
measurement accuracy. A 0.0001 in. (0.0025 mm) accuracy requirement over 10 in. (250 mm) corresponds to 2 arc seconds
of collimation. Since variations in the refractive index of air
due to temperature differences are capable of deflecting a light
beam 5 to 6 arc seconds, individual measurements can fluctuate
several micrometers, approximately 0.0003 in. (0.0075 mm) due
to atmospheric and thermal nonuniformities. Averaging will, of
course, substantially reduce this source of error. However,
since turbulence has a $1/f$ noise spectrum, averaging will reduce the error only a limited amount.

Dirt, dust, and oil are a reality of the manufacturing floor.
If these contaminants are in the measurement region in the form
of particles, droplets, or other beam-interrupting substances,
they will modulate the light beam and introduce uncertainty in
the measurements. As with other random errors, averaging
will reduce these effects. An oil film or particulate matter on
the object being measured can introduce systematic errors.
The magnitude of the errors depends on the size and nature of
the particular contaminant. An air wipe prior to measurement
can virtually eliminate this error.

The temperature of the part being measured can affect its
size. Since all engineering materials have some level of sensitivity to temperature, there will be a systematic error given by

$$e_T = d_{T_0} (a_p - a_i)(T - T_0)$$
(3.3)

where d_{T_0} is the dimension at standard temperature T_0, a_p is

the coefficient of thermal expansion of the part material, a_i is
the coefficient of expansion of the instrument, and T is the
temperature at the time the measurement is taken. Obviously,
if both coefficients of thermal expansion are well known, this
error can be corrected.

If the object being measured has a surface finish which has
roughness as large as the measurement beam diameter, the dimension of the part will have an error that is associated with
the microfinish of the part. The exact value of that error depends on the type of surface finish on the object. One type
of surface finish is that produced by turning the part on a
lathe. This pattern is similar to the threads of a screw. In
this case the error is given by

$$e_L = \frac{p}{2} \tan\left(\frac{d}{2}\right) \tag{3.4}$$

where p is the feed per revolution of the lathe cutter and d is
the angle of the cutting tool point. Other more random surface
finishes present different errors. The exact amount that the
LTS senses is a function of the beam diameter and the part
roughness profile as shown in Figure 3.3. The LTS has the
inherent capability to sense a measurement near the RMS surface of the object being measured. This is important to consider when comparing the LTS data with data obtained with a
contacting gauge, since contact gauges tend to measure peak-
to-peak dimensions.

Edge-sensing errors are known to occur on transparent objects, with the greatest effect seen on hollow tubes. The amount
of the error is affected by the part diameter, the laser beam diameter, and the wall thickness if the part is tubular. Typically
a transparent object will produce a signal with well-defined outer
edges, poorly defined inner edges (again, if the part is tubular),
and many other edges caused by reflection, refraction, dirt, or
interference as the light passes through the part. The extra
edges may be ignored through the use of special software or
hardware which inhibits the signal during noisy intervals. Under
some conditions the outer edges of the part do not produce a
satisfactory edge transition, which may result in an inaccurate
measurement. There are a number of theories which explain
the source of the error. One is that the wall thickness, in the
case of the tubular object, is less than the beam diameter, causing the signal from the outer edges to be distorted by the signal

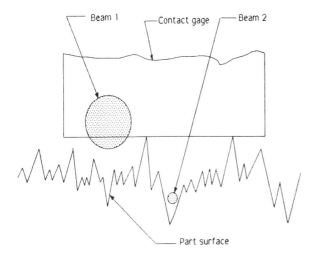

FIGURE 3.3   Error due to variation in surface finish and beam diameter.

from the inner edges.   Another theory is that a portion of the beam is refracted and/or reflected through the part with a small enough angle that it can still enter the collector lens and recombine with the actual edge signal, thus causing an error.   Generally, an instrument with a smaller beam diameter will produce the most accurate results, but each application must be reviewed to determine measurement limitations.

Stray light from external sources does not usually cause errors unless it is great enough to saturate the preamplifier or the photodetector.   Stray light will also cause a problem if the intensity variation is at a high enough frequency to pass through the differentiators.   An effective method for controlling stray light is to filter it out using an optical bandpass filter which transmits only the frequency of the laser light.   This is the method used in applications when measuring hot, incandescent objects.

Process errors are dependent on the particular process and cannot be generalized.   A particular process may produce conditions in which any or all of the error sources mentioned are present.   Other error sources could be produced by the introduction of additional optics in the laser beam path or by unique conditions of the process.   Special considerations are required if a possible error is suspected.

Internal error sources can be grouped into three categories: optical error sources, mechanical error sources, and electronic error sources. A fourth internal error source which appears in the newer instruments is the software error source.

The optical error sources in the LTS can be controlled by proper specification of components and fabrication processes. The precise reasons for these errors are not fully understood but enough empirical information has been gathered to make intelligent predictions and recommendations for the components in question. The laser has some inherent characteristics which cause errors in the LTS. These are the laser beam amplitude noise, beam pointing change, spot shape changes, and sensitivity to optical feedback.

Amplitude noise on the laser beam at low frequencies will not affect the measurement because of the high-frequency response of the differentiation electronics. Amplitude noise at higher frequencies may affect the measurement. This noise may be random or systematic. Random noise, as stated before, can be reduced by averaging. Systematic noise may be in the 1 to 2 MHz range, caused by electrical oscillation of the laser tube due to too small a ballast resistance, or in the 20 kHz range, caused by insufficient filtering on the laser power supply output. Systematic noise may affect accuracy by "pulling" a measurement slightly to be a multiple of the noise period. Systematic noise may also cause dimensions that are exact multiples of the noise period to appear more stable than other dimensions. The amount of possible measurement error can be determined by

$$e_m = \frac{kNV_b}{s_r} \tag{3.5}$$

where e_m is the measurement error, k is a scaling constant, N is the laser noise, V_b is the beam velocity, and s_r is the slew rate of the second derivative.

When the laser tube changes length due to temperature variations, small changes in the beam direction occur as the laser oscillation finds its most stable axis in the laser cavity. This effect occurs when the laser "modes" and causes the beam to strike the optics at a different position.

Optical feedback occurs when a reflection from one of the optical surfaces external to the laser is of sufficient intensity and is well enough aligned with the laser cavity to reenter the laser. This causes the laser beam intensity and pointing stability to become unstable and affect the measurement.

The output beam of a typical laser usually has a Gaussian energy distribution known as TEM_{00}. If the laser end mirrors are misaligned or some other nonuniformity exists in the laser resonator, the energy distribution could be different and cause errors in the measurement.

The solution to the first three error sources was to choose a polarized laser with good stability and inspect to assure acceptable noise levels at incoming inspection. The solution to the last error was to mount all optics near the laser with sufficient tilt to offset the back reflection from reentering the laser. Antireflection coatings on the optics were also tried, with limited success.

The error sources related to the collimating lens are as follows: ray slope errors, high-frequency slope errors, internal interference effects, cosmetic surface quality errors, scan speed linearity errors, and thermal expansion errors. Ray slope errors, thermal expansion errors, and scan speed linearity errors can be optimized in the optical design of the lens. The ability of the lens design to correct for these errors is limited by the type of lens or mirror chosen. As an example, the lens used in the first LTS was an air-spaced doublet which was designed to give ray slope errors of less than 2 arc seconds and constant scan speed (K-theta) performance. This lens was used with a simple electronics package that had no stored correction tables to maintain accuracy. The parabolic mirror used in the later LTSs has perfect collimation but requires a stored correction table to correct the scan speed linearity errors inherent in the design. The amount of ray slope error which can be tolerated in a scan lens design is given by

$$R_m = \tan^{-1}\left(\frac{e}{2M_T}\right) \tag{3.6}$$

where R_m is the maximum ray slope error in the lens, e is the desired measurement error of the system, and M_T is the measurement throat length of the system.

Internal interference effects are also design related and can
be avoided by following a few guidelines; all lenses should be
used with the optical axis tilted perpendicular to the scan direction with an angle sufficient to displace the beam one beam diameter. If tilting is not possible, a high-efficiency antireflection coating should be used. In any cemented lens design the
index of refraction of the cement should match the index of refraction of one of the elements. High-frequency slope errors
and cosmetic surface quality are fabrication process related.
Optics used in the LTS should have the final polishing done on
low-speed polishing machines since high-speed polishers produce
high-frequency slope errors. Cosmetic surface quality errors
are the result of defects in the surface of the optics in the
system. The operation of the LTS is based on the intensity
variation of the laser as it encounters an object. For this reason, any defect on the optics or dirt particle which causes an
intensity variation of the light signal can cause an error in the
measurement. Since the laser beam is small in diameter on the
optics in the system, a defect of 0.0012 in. (0.03 mm) in diameter can cause a significant error in the measurement. The allowable defect size for a given instrument can be approximated by

$$S_L = 2e \frac{d_o}{d_p} \qquad \text{for lenses} \qquad (3.7)$$

$$S_M = \frac{e(d_o/d_p)}{2} \qquad \text{for mirrors} \qquad (3.8)$$

$$D_L = 2\left(\frac{e}{d_p}\right)^{1/2} d_o \qquad \text{for lenses} \qquad (3.9)$$

$$D_M = \frac{(e/d_p)^{1/2} d_o}{2} \qquad \text{for mirrors} \qquad (3.10)$$

where S_L, S_M, D_L, and D_M are scratch and dig sizes for lens
and mirror surfaces, respectively, d_o is the laser beam diameter
on the optic in question, d_p is the laser beam diameter at the
passline, and e is the desired accuracy error for the instrument.
One can see that mirror surfaces are approximately four times
more sensitive than lens surfaces to defects. Also, these equations show that using a large beam on the optic (d_o) and a
small beam at the passline (d_p) increases the allowable scratch
or dig size for a constant error. The only way to control the

cosmetic surface quality is through tight quality control and
in-process inspection.

The windows are necessary to protect the internal components of the LTS from environmental contaminants and air turbulence. The windows on an LTS are subjected to the same constraints as the other optical components of the system. The
cosmetic surface quality, high-frequency slope errors, and mounting considerations used in the fabrication and assembly of the
windows are the same with one important addition. The internal
wedge of the windows causes an interference phenomenon that
modulates the output beam of the instrument. By tilting the
window and fabricating it with the wedge perpendicular to the
scanned beam, this error can be eliminated. The window used
on most photodetectors also is subject to internal interference
and can cause similar errors. The solution to this was simply
to remove the window from the photodetector. Another error
source to which the windows are subjected is environmental contamination. Collection of dirt, dust, and oil on the windows will
lead to degradation of accuracy in the instrument. Periodic
cleaning, improved environment, or special air wipes can reduce
or eliminate this error.

The collector lens is the least critical optical element in the
measurement beam of the LTS. The errors in the collector lens
which can cause errors in the accuracy of the LTS are cosmetic
surface quality, internal interference, and lens aberrations.
The topics of cosmetic surface quality and internal interference
have been covered in previous sections. Aberrations in the collector lens can cause two distinct errors in the LTS. The function of the collector lens, as the name implies, is to collect the
light and focus it onto the photodetector. Since the photodetector active area is usually quite large, the collector lens should
only need to produce a focused spot smaller than this area in
order for the photodiode to "see" the entire scanned beam. However, as the diameter of the part increases the amount of reflected light from the edge of the part which can enter the receiver increases. The error due to reflection off the part is
given by

$$e_R = d\left(\frac{1 - \cos\ g}{2}\right) \tag{3.11}$$

Here e_R is the error due to reflection, d is the diameter of the
part, and g is the field angle of the receiver. The error

introduced by this effect can be enhanced by the aberrations in the collector lens. A good example of this is a simple plano-convex lens as the collector. The positive spherical aberration in this lens at the point of minimum blur will cause the edge rays to have an effective field angle greater than the rays at the inner portions of the lens. This will increase the error given by Eq. 3.11. By focusing this system slightly ahead of the "best" focal position normally chosen, the errors can be reduced. By improving the lens design performance to reduce the aberrations present, the error can also be reduced.

The source of air turbulence internal to the LTS is some-what different from the external sources mentioned earlier; how-ever, the ray pointing errors are essentially the same. The external air turbulence is due primarily to the environment and is sometimes impossible to correct. There are two sources of internal air turbulence. One source is the temperature gradients and subsequent convection heat transfer which occur between the internal components of the LTS. The second source is the pumping action caused by the scanner as it rotates. Since the nature and location of the heat-producing components are well known, steps can be taken to minimize their effects. Also, the pumping action of the scanner can be isolated. The main heat sources in an LTS are the laser, the laser power supply, the scanner motor, and the internal electronics. The heat generated by these components can be isolated from the main optical path by careful layout in the design of the instrument. This removes the direct heat gain to the beam path. The indirect heating remaining can best be controlled by baffling the optical path to reduce the convection heat transfer. The circulation-reducing baffle (U.S. Patent 4,427,296) accomplished this in the following way. In order to reduce the heat transfer to purely conduction and radiation, air motion must be prevented. By designing a baffle system with the proper plate spacing, the air motion can be significantly reduced. An equation which gives the correct dimensions for such a baffle is

$$d_p = \left(\frac{3HTv^2}{g \ Pr \ dT} \right)^{1/4} \tag{3.12}$$

where d_p is the plate spacing, H is the height of the baffle plate, T is the nominal temperature at which the fluid properties are evaluated, v is the viscosity of the fluid (air in this case), g is the gravitational acceleration, Pr is the Prandtl

number, and dT is the desired temperature difference between
the plates. Baffles designed using Eq. 3.12 have successfully
reduced the system noise by at least a factor of 5.

The scanners used in the LTS have evolved over the years
in a variety of forms. Monolithic prisms with four or five sides,
cemented glass/metal assemblies, and two-sided glass plates have
all been utilized with good results. With all these designs there
is a certain amount of air motion created by the rotating prism.
This air motion, like the motion caused by the convection heat
transfer, causes errors in the measurement. The exact amount
of air turbulence created by the scanner is a function of the
diameter, the rotational speed, the number of facets, the construction, and the surrounding structure. Effective scanner
shrouds have been fabricated by placing an enclosure with a
small opening for the laser beam to enter and exit around the
scanner.

Mechanical error sources can be grouped into three categories: temperature-induced effects, stress-induced effects, and
vibrational effects. The transmitter optical bar is the area of
major concern when considering the mechanical error sources
and methods of reducing them.

Temperature-induced effects are by far the most difficult
error sources to correct. As mentioned earlier, all engineering
materials have some temperature sensitivity in that they have a
measurable coefficient of thermal expansion. A few are less
sensitive, but they can be expensive and difficult to work with.
Some examples are Invar, Zerodur, U.L.E., and fused silica.
Invar is an iron-nickel alloy containing 36 wt. % nickel, Zerodur
and U.L.E. are glass-ceramic alloys with proprietary composition,
and fused silica is synthetic quartz. There are composite materials which are insensitive to temperature but are sensitive to
humidity. Also, all optical materials used for visible light have
a refractive index thermal expansion coefficient. The dimensional stability of the LTS is derived from the ability of the
system to locate accurately the edges of a part in time and reference them to a known standard. To accomplish this, the optical path must remain constant from the time of manufacture to
the time the measurement is taken. Also, since the operating
temperature of the instrument can vary, the temperature sensitivity of the instrument must be held within some limits.

Two methods are used in a design which reduces the temperature sensitivity of the LTS optical bar. The first method
is to match various materials, usually a material with a low coefficient of expansion and a material with a high coefficient, to
produce an optical bar that has a temperature coefficient equal
to that of the collimating optic. In equation form, this is

$$\frac{d(OPL)}{dT} = \frac{d(OBL)}{dT} \tag{3.13}$$

where d(OPL)/dT is the change in the optical path length of
the collimating optic due to temperature and d(OBL)/dT is the
change in optical bar length due to temperature. These two
quantities are given by

$$\frac{d(OPL)}{dT} = fa_{eff} \tag{3.14}$$

$$\frac{d(OBL)}{dT} = \{L_1 a_1 + L_2 a_2 + \cdots + L_n a_n\} \tag{3.15}$$

where f is the focal length of the collimating optic; a_{eff} is the
effective coefficient of thermal expansion for that optic, which
includes dimensional and refractive index changes due to temperature; L_1, L_2, ..., L_n are the lengths of the optical bar
components; and a_1, a_2, ..., a_n are the coefficients of thermal
expansion for the respective optical bar elements. This approach
"athermalizes" the optical bar and is used for instruments which
require a large depth of measurement range or high accuracy.

The second method utilizes the fact that a ray entering a
lens at a given angle maintains the same ray height at the front
focus of the lens for small changes of the origin of that ray
along the optical axis. This limits the position where the instrument can accurately measure but allows the designer to simplify
the optical bar considerably. This also allows the use of materials with higher coefficients of expansion, such as aluminum,
for the optical bar without any adverse effects.

Other thermally induced errors are related to differential
expansion of the optical mounts and the optics. There are
metals, such as Kovar, which are intended to match thermally
some optical materials, but the selection is limited and cost or
fabrication may present problems. This forces the use of common materials with special mounting techniques. One widely
used method of optical mounting is to use an elastic adhesive

such as RTV silicone rubber. RTV allows differential expansion
to occur but holds the optic securely to its locating surface during use. Another technique clamps the optic in pure compression with resilient pads. There is some risk, since the mounting surfaces must match exactly to guarantee that no bending
moments are induced. A third method athermalizes the optical
mount by calculating the amount of adhesive thickness necessary
to compensate for the differential growth of the mount and optic.

Stress-induced error sources refer to any built-up stress
in the LTS which is not due to differential thermal expansion.
This includes assembly stresses induced by dimensional errors
in components and tightening of fasteners, weight-induced
stresses, and mounting-induced stresses. Assembly and mounting-induced stress can best be controlled by kinematic design
and designs which allow for component inaccuracies.

Vibrational errors can be induced by the motor scanner
assembly. These usually occur if the scanner is not balanced
and another component in the system has a natural frequency
near the rotational frequency or drive frequency of the motor.

Electrical error sources can be grouped into two categories:
pure electrical errors and electro-opto-mechanical errors. The
pure electrical error sources are preamp electrical noise, insufficient preamplifier bandwidth, signal quality degradation, signal
distortion, group delay distortion, time interval measurement
errors, offset voltage errors, and response time errors. The
electro-opto-mechanical error sources are motor speed errors
and photodetector response nonuniformity.

Preamplifier noise sources are shot noise from detector dark
current and op-amp bias current, Johnson noise from detector
source resistance, feedback resistor or detector load resistor,
amplifier short-circuit noise voltage, and external noise. Amplifier short-circuit noise is usually the largest contributor,
since the detector capacitance and feedback resistance combination causes it to be amplified at higher frequencies. External
noise can also get into the signal path from the power supply
and from capacitive coupling to the circuitry or detector. Noise
from the power supply is easily reduced with proper regulation
and filtering. Capacitively coupled noise may be reduced by
connecting a capacitor from circuit ground to chassis ground
near the preamp. The preamplifier design should be optimized

for low noise over the entire bandwidth of interest. Usually
the signal from the detector is great enough that noise is not
a problem.

The bandwidth required for the preamp and signal processing path depends on the laser beam spot size on the part and
the beam velocity. The electrical signal of the second derivative has its spectral peak at a frequency given by

$$f_d = \frac{4V_b}{d_p} \qquad\qquad (3.16)$$

where f_d is the spectral peak frequency and V_b and d_p are
the beam velocity and beam diameter, respectively. The electronics should ideally have a bandwidth about three times this
frequency. Since the noise also increases with bandwidth, the
actual bandwidth is usually only one to two times this frequency.

If the electronics is partitioned with the preamp in a separate enclosure from the other electronics, proper cable driving
and termination practices should be observed so that the signal
quality does not depend on cable length. It may also be difficult to prevent external noise from adding to the signal on the
cable. If the preamp, differentiators, and zero-crossing detector are in a separate enclosure from the other electronics, digital signals may be sent through the cabling with less worry
about noise and signal degradation. The associated logic should
be designed so that a pulse on a single wire is produced for any
edge transition. If this is not done, variations in logic threshold voltage or device propagation delay may affect the measurement.

Signal distortion usually falls into two categories, slew rate
limiting and group delay distortion. Slew rate limiting occurs
when the signal changes more rapidly than the amplifier can
follow. This is distinctly different from bandwidth limitation
and may be recognized by the fact that rising and falling edges
of the signal have constant slope rather than smooth curves.
Group delay is defined as the derivative of phase of the transfer function. Group delay distortion occurs when some frequency components of the signal are delayed more than others
as they pass through the electronics. When the scanned laser
beam is focused to a smaller spot in the measurement region,
the line of sharpest focus is slightly curved. Therefore, the

frequency spectrum produced by one edge of the object being
measured may be slightly different from the spectrum produced
by the other edge. If some frequencies are delayed more than
others in the electronics, an apparent measurement error may
occur.

The resolution of the time interval measurement must be
finer than the random errors from other sources. If the random error sources are greater than the quantization uncertainty,
both may be reduced together by averaging. If the random
error sources are less than the quantization uncertainty, the
result of averaging will still contain the quantization errors.
This results in the paradox that reducing the random errors
from other sources below this level might actually result in less
accurate measurements after averaging.

The time interval error source depends on the counting
resolution of the reference clock. Direct counting of time intervals by counting clock cycles becomes difficult above 40 MHz
(25 ns). A typical high-accuracy LTS system with 10.0×10^{-6}
in. (0.25 micrometer) repeatability requires 1.1 ns counting
ability. To obtain finer resolution, two methods are most attractive: Delay line interpolation (U.S. Patent 4,332,475) passes
the edge signal through a tapped delay line with N taps and a
total delay equal to the reference clock period. Each cycle of
the reference clock takes a snapshot of the N delayed edge signal to determine a fractional time interval. This approach has
a practical resolution limit of about 5 ns. Analog interpolation
is the second approach. This technique uses a pulse which is
generated with a width of one clock period plus the unknown
fraction. A constant current is integrated during the width of
this pulse, and the resulting voltage is measured. This approach
is more complicated but has a practical resolution limit of 0.5 to
1.0 ns.

Offset voltage errors occur if the zero-crossing detector
for the second derivative does not switch at exactly zero volts.
This effect can be seen when the two edge transitions are in
opposite directions, as for a hole or a diameter measurement.
The dimension will be measured either smaller or larger than
it actually is, depending on the polarity of the error. The
amount of error may be calculated in a manner similar to the
calculation shown for laser amplitude noise (Eq. 3.5). This
error will cancel out if the direction of edge transition is the
same on both edges.

Errors due to response time may be present since most comparators have a response time that is affected by the slope of
the input signal, the direction of the slope of the input signal,
temperature, and load on the output. To reduce this error a
comparator must be chosen which has a fast, predictable
response.

Motor speed errors can be any or all of the following: low-
frequency (average) motor speed changes, high-frequency (instantaneous) motor speed changes, on-off effect, and electromechanical vibrational coupling.

Since the beam velocity is directly proportional to the motor
velocity, motor speed errors directly affect the accuracy of the
instrument. Two types of motor speed errors are present in
any electric motor. They are low-frequency or average speed
fluctuations and high-frequency or instantaneous speed variations.

Low-frequency motor speed changes cause the measured
size of a part to fluctuate inversely with the speed of the motor.
This can be corrected by ratioing a stored reference edge dimension with the apparent reference edge dimension and correcting
the part size accordingly. In this system a synchronous motor
with a high rotational inertia can be used, powered by the AC
line. The ideal motor for this system is a brushless DC motor
with precision bearings, a large inertial mass, and a simple
speed control, similar to the motors used in small hard disk
drives. In a system without a reference aperture the motor
speed must be derived from the same clock used to measure the
time intervals. A synchronous motor operated on power derived
from the measurement clock is one possibility. This has the
disadvantage of requiring a power amplifier to provide typically
10 watts of power to the motor. The ideal motor for this system may be a brushless DC motor with a very accurate, high-
resolution encoder and phase-locked speed control system.
However, this tends to be rather expensive.

High-frequency speed variations are caused by the effects
of motor pole effects or feedback loop effects in the case of a
DC motor. High-frequency speed changes cannot be removed;
instead the motor needs to have enough inertia to reduce this
effect to an acceptable level. The large rotational inertia im-
proves both low- and high-frequency motor speed fluctuations.

Another method of reducing the effects of high-frequency motor speed errors is to average the measurement data in multiples of the number of scanner facets. This guarantees that each measurement is the average for one or more complete revolutions of the motor.

The error induced by turning the instrument off and on is very unusual and was discovered in testing the early systems. In the first prototype instruments the scanners were designed to be four-sided glass cubes. When the units were being tested a large repeatable error was observed when the unit was turned off and back on. After much further testing the problem was found to be related to the high-frequency speed variations of the motor and the number of facets on the scanner. The synchronous motor being used would synchronize on a different pole each time the motor was turned on. This caused the speed profile over each scan to be different each time the unit was powered up. The effect was also found to occur only when the number of poles in the motor was an even multiple of the number of facets on the scanner. The solution to the problem was simply to increase the number of scanner facets to five. The five-sided scanner produced scan speed fluctuations which were random for any pole position of the motor, so when the measurement data were averaged the on-off error was eliminated.

Electromechanical vibrational coupling occurs most often in systems in which the motor is driven off the AC line voltage. This error is caused by resonant vibrations in the rotor, scanner, and electrical windings as the motor is run at different frequencies. The resonant vibrations of the rotor-scanner assembly beat against the motor poles and cause changes in the high-frequency motor errors as a function of drive frequency. This error can be reduced by reducing the scanner inertia or increasing the motor shaft diameter to raise the natural frequency well above the motor drive frequency. It is important to note that a 0.25-micrometer error can correspond to a motor angle change of as little as 0.2 microradians.

As the laser beam scans, the point where it is focused on the receiver diode moves slightly. If the detector has nonuniform response over its active area, motion of the spot on the detector surface will appear to be a change in signal. This can produce confusing symptoms similar to those caused by optical interference in the windows.

Math errors in software can occur on any system in which calculations are done by a programmed routine. The more complicated the system, the higher the likelihood of "bugs" and the more difficult it will be to find them. By careful study of the software by more than one person and by statistical distributions, these bugs can usually be found and corrected.

The other components of an LTS system necessary for a complete instrument are the scanner, the beam-shaping optics, and the sync. detector or auto-cal aperture.

Figure 3.9 shows a modern system schematic of an LTS. This system incorporates the additional components which are common in a modern LTS system.

The scanner was mentioned earlier as evolving over the years in various forms. Two-, four-, and five-faceted scanners have been used with success. Both single-piece glass scanners and composite scanners of metal and glass are in use today. The reasons for the various scanner types are many. Two reasons already mentioned were electromechanical vibration and on-off motor pole effects. Two other reasons for the different types of scanners are the duty cycle and number of scans required for a given time interval. It is obvious that for a given system the percentage of time actually used for an active scan during one revolution of the motor can be doubled simply by doubling the number of facets on the scanner. Also, if the system is electronically limited to a maximum beam velocity, the only way to increase the number of scans per second is to increase the number of facets on the scanner.

The additional requirements of the scanner explain the composite versus solid scanners. The other properties important to the scanner are the angular pointing error between facets, the distance of the scanner facet from the best-fit cylinder, and the figure and cosmetic surface quality of the facet surfaces. Angular pointing errors between facets on a scanner refer to tilt of a scanner facet perpendicular to the scan direction. This error can be seen if the scanned beam is viewed at a great distance from the scanner and separate scan lines are present. In the LTS, this causes the scans from different scanners to hit the other optics and auto-cal at different places, which may produce errors.

If the distances of the scanner facets to the center of rotation of the scanner are not equal, there will be a slight focus
error from facet to facet as the system operates. The errors
caused by this effect will vary for each system and may cause
an increase in the random noise of the measurement.

The scanner is as sensitive to cosmetic defects and surface
figure as the other components in the transmitter. In some
cases the scanner is near an intermediate beam waist and becomes even more critical.

These stringent requirements for the scanner made it necessary to use only optical polishing to produce the required
surface for the LTS. Because of process control and inspection
limitations 5 to 10 years ago it was more economical to build composite scanners from glass mirrors bonded to metal hubs. Even
now the technology for fabricating monolithic scanners with the
desired performance is still being developed, so the replacement
of composite assemblies is slow.

The beam-shaping optics are used to expand or focus the
laser beam to the desired size for a particular instrument or
application. As mentioned in the section on surface roughness,
the beam size plays an important role in determining the sensitivity of the LTS system to part roughness. Also, the beam
size affects the ability of the electronics to sense an edge accurately as well as the measurement throat range of the instrument and the sensitivity of the instrument to cosmetic defects
on the optics. A small beam increases the slow rate of the second derivative and decreases the zero-crossing detection error.
A small beam also diverges faster as given by

$$w(z) = w_0 \left[1 + \left(\frac{\lambda z}{\pi w_0^2} \right)^2 \right]^{1/2} \tag{3.17}$$

where $w(z)$ is the beam radius at distance z, w_0 is the beam
waist radius, λ is the wavelength of the laser light, and $\pi =$
3.1416. Since the gain of the electronics is optimized at the
passline of the instrument, the signal degrades as the beam
size increases. The beam size is usually limited to $w(z) =$
$1.414 w_0$. This criterion forces instruments with long measurement throat requirements to have large-diameter beams and
high-accuracy instruments to have small-diameter beams. The
types of beam-shaping optics vary from single-element lenses

to two-element telescopes which can expand or contract the
beam and can be set to various effective focal lengths. The
beam waist size and position in the measurement region can be
calculated by consecutive iterations of the following equations:

$$\frac{1}{w_2^{\,2}} = \frac{1}{w_1^{\,2}(1 - d_1/f)^2} + \left(\frac{1}{f^2}\right)\left(\frac{\pi w_1}{\lambda}\right)^2 \tag{3.18}$$

$$d_2 = (d_1 - f)\left[\frac{f^2}{(d_1 - f)^2} + \left(\frac{\pi w_1^{\,2}}{\lambda}\right)^2\right] \tag{3.19}$$

where d_1 and d_2 are the distances from the optic to the first
and second beam waists, respectively, f is the focal length of
the optic, λ is the wavelength of the laser light, $\pi = 3.1416$,
and w_1 and w_2 are the first and second beam waist radii,
respectively.

Beam expanders can be made using cylindrical lenses to expand the beam in only one direction. Instruments with these
beam expanders have been used to measure extremely rough
surfaces.

The sync. detector is used to trigger the electronics to
start looking for measurement edge data. It is the method used
on the early systems and relies on a constant motor speed for
accuracy. Auto-calibration (auto-cal) replaced the sync. detector on later systems and provided a reference aperture to be
used to correct for low-frequency motor speed fluctuations,
which increased system accuracy. Also, the auto-cal can correct for inaccuracies in the athermalization of the optical system.
The auto-cal can be placed in a number of different places with
little effect on its function. The auto-cal consists of a mask
with two reference edges or slots with a set of photodiodes behind each edge or slot. Whether an edge or a slot is used is
determined by the electronics of the system, but the function
remains the same. The time between reference pulses is measured on each scan and ratioed with a stored value. The object measurement is then multiplied by this ratio to obtain the
correct output dimension. The material used for the reference
mask depends on the system configuration. A system which is
athermalized will require a mask with low thermal expansion,
while a system working at one focal length will require a mask
matched to the thermal expansion of the optical bar material.
The most critical feature of the auto-cal mask is the parallelism

and surface finish of the reference edges or slots. Any non-
parallelism or surface roughness may be seen by the instrument
since temperature or scanner errors may move the beam along the
auto-cal. The apparent change in the reference aperture will
cause the instrument to output an error in the measurement.

In summary, the design of a new LTS is developed by considering all the error sources that have been discussed in this
section and making calculations and judgments based on them.
It must be realized that this information has been compiled over
more than 13 years of experience and testing.

The remainder of the chapter describes the embodiment of
the design ideas and constraints. Examples are given with design details and performance specifications of LTS units manufactured by the Zygo Corporation.

MODELS 110, 120, AND 121

Figure 3.1 shows schematically and pictorially the original LTS
system. It featured separable transmitter and receiver, 2 in.
(50 mm) maximum part size, and 10 in. (254 mm) measurement
throat. The system accuracy was 0.0002 in. (0.005 mm). The
system used a sync. detector to trigger a measurement and a
K-theta lens and hysteresis synchronous motor to provide stable
accuracy without a stored correction table. The scanner in
this unit was five-sided to reduce the on-off effect inherent in
the motor. The optical bar in the transmitter was athermalized
using Invar rods with aluminum end plates. The model 120 was
the next-generation unit. It was essentially the same as the
110, with updated electronics which used stored correction values
to improve accuracy. The model 121 was the final version of this
instrument. The K-theta lens was replaced in the standard unit
at the sacrifice of the 10-in. measurement throat. The 121 also
had a stored correction table. By eliminating the K-theta lens,
the cost of the instrument was reduced. The beam shaping on
these units was done by passing the laser beam through the
collimating lens before it hit the scanner. This gave an output
beam with a large beam waist which allowed the large measurement throat. The collecting lens in all three units was a single-
element lens. There were also focused beam units in this family
(120F, 121F) which had an additional lens in front of the laser.
The reduced beam size improved the accuracy but limited the
measurement range and throat to only 0.5 in. (13 mm) and
0.125 in. (6 mm), respectively.

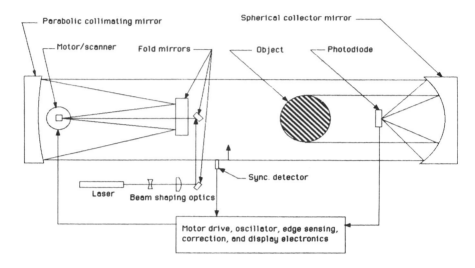

FIGURE 3.4   Simplified schematic model 130/131.

MODELS 130 AND 131

Figure 3.4 illustrates the first 4 in. (100 mm) aperture system,
the model 130.   This unit featured the same separable transmitter and receiver, with a 4.5 in. (114 mm) measurement throat
and a system accuracy of 0.0005 in. (0.013 mm).   The collimating optic in this system was a parabolic mirror, which used
a stored correction table and sync. detector.   The optical bar
was athermalized with Invar and aluminum as in the 2-in. units.
The model 131 was essentially the same instrument with some
cost-saving measures.   These systems utilized a two-element
beam expander to produce an output beam which had a 0.04 in.
(2 mm) diameter.   This allowed the large throat length.   The
two units used a sync. detector and a synchronous motor driven
off the internal clock.   Both the 130 and 131 used a spherical
mirror in the receiver and a five-sided scanner.

MODELS 1532 AND 1535

Figures 3.5a and b show the 1500 series LTS.   This instrument
represents the culmination of knowledge in the LTS field.   It

was the first to use baffling, analog interpolation, and a DC brushless motor. It also featured many special measurement types and process control functions as software options. U.S. Patent 4,427,296 was based on the inventions in this instrument. The correction table values were set using a laser distance measuring interferometer instead of gauge pins as in earlier units. The optical bar was athermalized using Invar, stainless steel, and aluminum. Because of the accuracy requirements of these systems, most of the errors mentioned in the previous section of this chapter became very apparent. The temperature effects proved extremely difficult to control in production because of the tolerance stackup of dimensional errors and material inconsistencies. Autocalibration and a scanner shroud were added later to improve the long-term stability and allow the use of a less complicated scanner motor system. The model 1532 has a 2 in. (50 mm) measurement range with an accuracy of 0.00003 in. (0.00076 mm), and the model 1535 has a 4 in. (100 mm) measurement with an accuracy of 0.00012 in. (0.003 mm). These systems have beam expanders which produce a small-diameter beam waist at the passline. Moreover, to ensure the accuracy

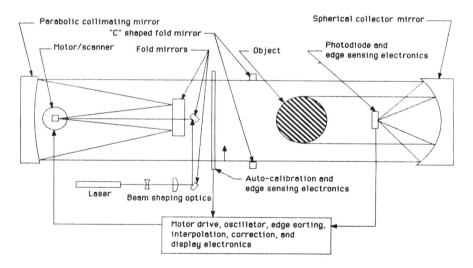

FIGURE 3.5a Simplified schematic models 1532 and 1535.

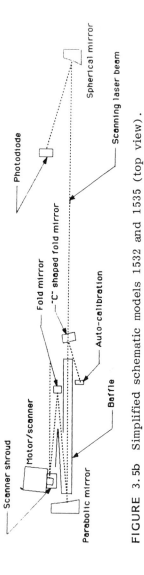

FIGURE 3.5b   Simplified schematic models 1532 and 1535 (top view).

of these instruments, the auto-cal system was placed at the virtual passline of the system.   This was done by placing a C-
shaped mirror in the scanned beam which reflected back only
the outermost portions of the top and bottom of the beam.   The
auto-cal was then placed at the same position as the part would
be, but inside the unit.   In this way, the electronics could take
full advantage of the small beam diameter in sensing the edge of
the reference.   The auto-cal mask was made from Zerodur, since
this glass-ceramic material has a coefficient of thermal expansion
that is near zero.   Both systems use a parabolic collimating mirror and a spherical receiver mirror.

MODEL 1537

The model 1537 is shown in Figure 3.6.   This instrument has
an 8 in. (200 mm) measurement range, a 2 in. (50 mm) measurement throat, and an accuracy of 0.0006 in. (0.015 mm).   The
optical bar in this system is also designed using Invar, stainless steel, and aluminum.   This system uses a spherical collimating mirror and a spherical collecting mirror.   The collimating
mirror was required to have a long focal length in order to use
a spherical surface while maintaining adequate collimation accuracy.   This dictated the use of folding mirrors so that the

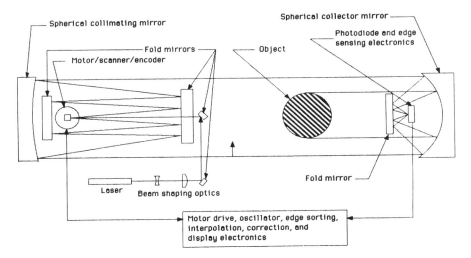

FIGURE 3.6 Simplified schematic model 1537.

package of the optical system had a reasonable size. The original design utilized a brushless DC motor with a precision encoder
and control circuitry. The more recent systems have auto-cal
and a brushless DC motor used in hard disk drives for cost reduction and reliability. The controller is the same as that used
in the 1532 or 1535.

MODEL 1561

The model 1561 is the largest monolithic LTS manufactured at
Zygo. This model is shown in Figure 3.7. The measurement
range of this unit is 18 in. (457 mm) with a measurement throat
of 8 in. (200 mm). The 1561 represents the most ambitious design used to date. The design was mainly driven by the constraint that the packaging size was to be as small as possible.
The collimating system is a telephoto system known as a cata-
dioptric telescope. This consists of a spherical primary mirror
and a mangin secondary. A mangin is a mirror with two spherical surfaces, of which the rear surface has a high-reflectivity
coating. This causes the mangin to act as both a lens and a
mirror. The effective focal length of the collimator is 90 in.,
while the overall length of the transmitter is only 33 in. The
optical bar was athermalized in the same way as the 2-, 4-,

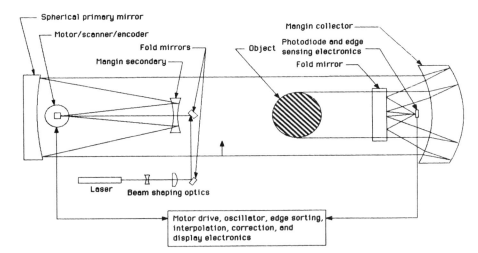

FIGURE 3.7 Simplified schematic model 1561.

and 8-inch units. The collector mirror is also a mangin and is
used double pass with the addition of a fold mirror. This design gives a 17-in. focal length with an 18-in. aperture. The
rest of the system is the same as in the other 1500 series LTS
units.

MODELS 1201 AND 1202

The 1200 series are a departure from the original Zygo LTSs in
that they are self-contained bench gauges. These instruments
have the transmitter, receiver, and processing electronics in a
single package. The primary design goal in this series was to
produce a low-cost, accurate bench gauge to compete with the
other scanning laser gauges on the market. Figure 3.8 shows
the schematic for the 1200 series. The single-enclosure design,
fixed measurement area, and low-cost requirements produced a
different optical bar design philosophy. Therefore, the center
of the measurement area was placed at the front focal plane of
the collimator. The thermal stability of the instrument was now
based on an optical phenomenon and not on engineering design.
The optical bar could then be made of any convenient material,
in this case aluminum. The system was originally designed with

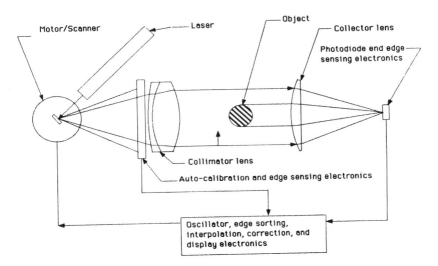

FIGURE 3.8 Simplified schematic models 1201 and 1202.

single-element lenses in both the transmitter and receiver in
the 1201, while the 1201 had an achromatic doublet lens as the
collimator. Unfortunately, the collimation errors (ray slope
errors) were too large and both systems were much too sensitive to object position for the gauge to be practical. Both systems were redesigned with higher-quality optics which had acceptable ray slope errors. An auto-cal reference aperture was used
to determine the timing reference for the system. This allowed
the use of a synchronous motor driven with AC line voltage.
Since the frequency of the line changes from 50 to 60 Hz across
the world, the electromechanical vibration of the motor scanner
assembly was a concern. The insensitivity of the final design
to frequency was verified by testing the unit over a range of
45 to 65 Hz. The measurement range, throat, and accuracy of
the 1201 are 2 in. (50 mm), 0.125 in. (3 mm), and 0.0001 in.
(0.0025 mm), respectively. Similarly, the 1202 performance
specifications are: measurement range, 1 in. (25 mm); measurement throat, 0.06 in. (1.5 mm); accuracy, 0.00006 in., (0.0015
mm).

MODELS 1101, 1102, AND 1104 AND
THE 1100 PROCESSOR

The 1101, 1102, and 1104 are the newest LTS systems in production at Zygo. Diagrams for the 1102 and 1104 are shown in
Figures 3.9 and 3.10, respectively. These systems are upgraded
designs of the early 121 and 131 models. The design goals for
the 1100 systems were to incorporate the advances in LTS technology into a more accurate and versatile version of the older
proven optical systems. These changes included auto-cal, baffling, and optical and mechanical design improvements. The
1101 and 1102 took the 121 optical bar layout, replaced the collimator and collector lenses with more efficient air-spaced doublet lenses, and added auto-cal, a baffle, and a scanner shroud.
The new collimator lens incorporated K-theta performance with a
shorter focal length. The new collector lens produced lower
aberrations with a shorter focal length. The mechanical improvements reduced assembly time and the total number of components
in the system and replaced the internally driven motor with a
less expensive line-driven one. The new instruments have a
smaller package size and better performance. The 1101 specifications are 1 in. (25 mm) measurement range, 0.5 in. (12 mm)
measurement throat, and 0.0008 in. (0.002 mm) accuracy. The

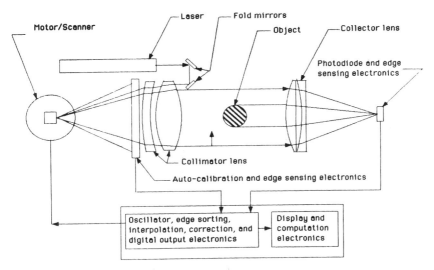

FIGURE 3.9   Simplified schematic model 1102.

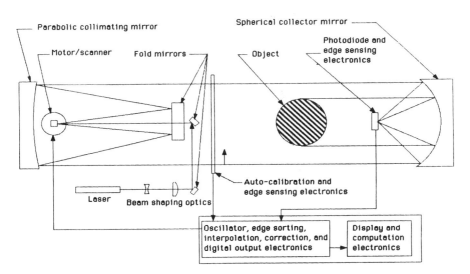

FIGURE 3.10   Simplified schematic model 1104.

1102 specifications are 2 in. (50 mm) measurement range, 1 in. (25 mm) measurement throat, and 0.00015 in. (0.0038 mm) accuracy. Similarly, the 1104 is the modernized version of the model 131. The auto-cal, scanner shroud, and line-driven scanner motor were incorporated in the 131 to make the 1104. Also, the mechanical components were improved to increase stability and reduce costs. The most notable change in the system was to replace the beam-shaping optics to produce a smaller beam size in the measurement region. This allowed the surface quality specifications of the optics to be relaxed, which further reduced the cost of component parts. The measurement range, measurement throat, and accuracy of the 1104 are 4.5 in. (114 mm), 2 in. (50 mm), and 0.0003 in. (0.0076 mm), respectively.

The 1100 controller is a new approach to the LTS instruments. All previous controllers were dedicated to one transmitter and receiver. The 1100 allows up to four transmitter and receiver pairs or sensors to be operated at one time. In addition, it can receive data from other sensors, such as temperature, velocity, and thickness sensors, over an RS232 interface. The data can be taken from all sensors simultaneously and can be manipulated by user-defined expressions or displayed. The 1100 incorporates most features, which were options on earlier LTSs, as standard. The 1100 can operate in such a way because each sensor has its own microprocessor for storing the correction table and basic gauge functions and the controller has its own microprocessor for the other display-related functions. Also, the standard software includes statistical process control, histogram functions, transparent object measurement, and many other features. These capabilities make the 1100 controller ideal for in-process and statistical process control applications.

CONCLUSION

Despite its simple concept, the LTS implementation was a significant technological challenge. The design of an LTS required the consideration of many aspects of electrical, optical, and mechanical engineering plus the characteristics of the components in order to produce the required accuracy and stability. The engineering design criteria are equally based on theoretical and empirical data determined through testing. The empirical data must be considered when designing new instruments so that improvements can be made.

4

In-Process Optical Gauging for Numerical Machine Tool Control

STEVEN SAHAJDAK / Consultant, Cleveland, Ohio

INTRODUCTION

During the past two decades numerical control technologies for
machine tools have advanced to the point where the limiting
factor is now the technique used for real-time measurement of
the controlled process (and the attendant inferred measurement
of the workpiece) in the feedback loop. All methods used today
to close the loop around the machine tool or the ball screws
leave machine deflections, tool wear, backlash, screw/nut errors,
thermal expansions, etc. outside the control/measurement loop.
This chapter deals with the study of loop closure by monitoring,
using a laser optical technique, the dimension of the actual workpiece as close as possible to the point of contact of the cutting
tool so that all errors due to the machine and tool are within
the control loop.

The major part of this chapter deals with the investigation
of a remote, noncontact optical gauging system required for an
in-process measurement by the machine control. The desired

This chapter was derived from the thesis of Steven Sahajdak.
The work was funded by the National Science Foundation under
grant DAR-7821639. The principal investigator was Professor
Harry Mergler at Case Western Reserve University.

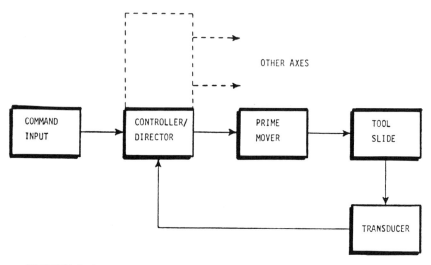

FIGURE 4.1   Basic numerically controlled machine tool loop.

capability of the gauging system is a measurement accuracy better than ±0.0005 in. at a standoff distance of 8 in. from the
machined surface. The optical gauging principle described exceeds this capability, and it can be adapted to a variety of measurement applications in machine control, automated assembly,
and robotic devices.

The optical gauging system was evaluated on an engine
lathe and the applications described pertain to the lathe but,
as can easily be inferred, the principle can be applied to a
variety of uses.

Any machine tool, although programmed or operated in a
manner to achieve a desired workpiece geometry, will introduce
errors into the geometry because of conditions such as backlash
and deflection of the machine. When cutting tough materials
these errors are certainly prevalent and in some cases exacer-
bated by work-hardening of the material in the near surface,
which introduces an additional constraint on achieving a desired
finish geometry. In order to compensate for these conditions
traditionally a trial-and-error approach is taken.

A basic numerical control loop is shown in Figure 4.1. Here
the command is applied to the controller, which interprets the

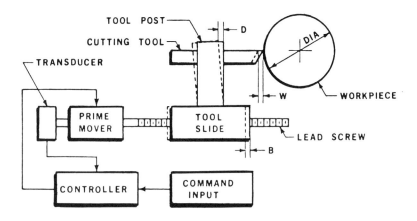

W—WEAR OR IMPROPER POSITIONING OF THE CUTTING TOOL
D—DEFLECTION OF THE TOOL POST UNDER VARYING LOADS
B—LEAD ERRORS AND BACKLASH IN THE LEAD SCREW

FIGURE 4.2 Conventional numerically controlled engine lathe and the resulting errors.

command and the feedback data to drive the slide by means of the prime mover. Feedback for the controlled dimension is obtained by monitoring the movement of the cutting tool relative to the workpiece with a transducer coupled to the tool slide or the ball screw. In this type of measurement scheme, the controlled parameter is not being measured, but instead another parameter which is not necessarily related to the first by a time-invariant one-to-one correspondence. This causes machining errors which cannot be compensated by the control system. For technical expediency, these errors are assumed to be constant and zero in magnitude. Furthermore, large computer numerical control (CNC) machines perform many operations with different tools, so precise and expensive approaches are taken to preset the tools to circumvent partially these errors.

Figure 4.2 shows a view along the workpiece axis using an example specific to the lathe. Command input is applied to the controller, which drives the machine tool slide via the prime mover. Feedback is obtained from a transducer coupled to the lead screw so that the diameter of the workpiece is determined by the position of the lead screw. Errors due to wear or mislocation of the cutting tool, deflection of the tool post under

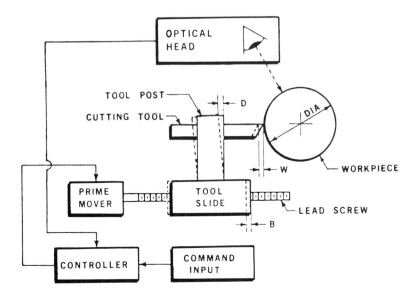

W—WEAR OR IMPROPER POSITIONING OF THE CUTTING TOOL
D—DEFLECTION OF THE TOOL POST UNDER VARYING LOADS
B—LEAD ERRORS AND BACKLASH IN THE LEAD SCREW

FIGURE 4.3   Numerically controlled engine lathe with optical gauging.

varying loads, and lead errors and backlash in the ball screw are all outside the control loop. Normally these errors are assumed to be constant and zero.

The ideal solution to machining errors in a CNC machine would be to close the control loop around the workpiece and obtain feedback for the controlled dimension by monitoring this dimension directly. This concept is shown in Figure 4.3. Here an optical gauging device is shown which can obtain information concerning the diameter of the workpiece directly. Ideally, this device should have a standoff distance on the order of 8 in. so that it could be mounted a safe distance away from the machining area. The area measured on the workpiece (spot size) should be small and well defined so as not to cause ambiguity when measuring nonplanar surfaces. Also, the device should provide an actual measurement of the dimension of concern rather than an indication of whether the measurement is

greater than or less than a preset size. Location of the measurement should be as close to the cutting tool as possible to
minimize the delay between the measurement and the feedback
to control the workpiece geometry.

As shown in Figure 4.3, control is applied in the same manner as in the previous example. The difference here is that
feedback for the diameter is obtained directly from the workpiece. Thus any of the errors previously discussed are now
within the control loop and are consequently eliminated by the
feedback. A unique advantage of this approach is that now
the tool need not be located precisely with respect to the workpiece; thus the expense and complications (cutter and holder
inventory, tool presetting equipment, personnel, etc.) of a
tool presetting system can be minimized or eliminated.

The optical gauging device, which is described in detail
later in this chapter, is described qualitatively here for an
overall understanding of the physical implementation of the
system. The optical gauging system is a noncontact distance-
measuring device which measures directly the distance from the
workpiece to a reference station. The distance between the
optical head and the measured surface can be scaled from less
than 0.4 in. to more than 20.0 in. and the parameters which
affect measurement accuracy are not directly related to this
standoff distance. Figure 4.4 shows the configuration of the
optical system on a lathe. The optical head should be positioned
so as to minimize the difference between the measured location
and the point of contact of the cutting tool to avoid additional
error. In practice, it may not be feasible to mount the head
at the cutting tool. In this application, the head was mounted
so that the point of measurement was directly opposite the cutting tool and well away from the machining area.

Due to the inevitable delay between the cutting operation
and the measurement of the workpiece, the feedback cannot compensate for rapid random backlash in the cross slide. Therefore, in this application a preloaded ball screw similar to that
found in a typical CNC machine was utilized. A stepper motor
was used to drive this axis to position the cutter. In a conventional feedback system, delay in feedback causes poor servo
response. In this system the incremental control of the tool
position permits the cutting tool to be moved by an increment
equal to the difference between the commanded radius and the

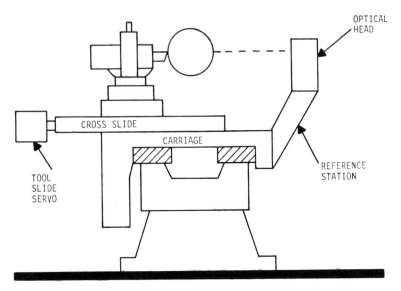

FIGURE 4.4 Practical configuration of the optical system on an engine lathe.

measured radius of the workpiece, which permits immediate compensation of machining errors after one delay period.

In this application on an engine lathe, the optical system measures the distance from the cylindrical workpiece to a reference station mounted on the carriage of the lathe. This mounting assumes that the axis of the workpiece is fixed relative to the carriage. Errors due to thermal rise of the spindle, deflection of any member of the machine, play in the carriage, and deflection of the workpiece are not included in the control loop because an indication of the radius, and not the diameter, is being measured. These errors can be eliminated with an additional optical head on a common reference station to obtain a true diametric measurement as shown in Figure 4.5. For the purpose of describing the optical gauge, which was the major effort of the research discussed here, a single-head configuration will suffice.

Since the optical gauge measures an opaque surface by reflected light, the measurement will be affected by liquid films on the surface of the workpiece. A correct measurement can

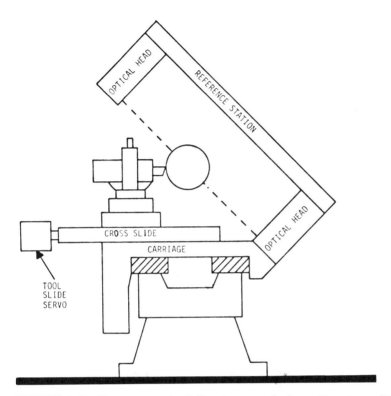

FIGURE 4.5 Measurement of the true workpiece diameter with the optical gauging system.

be restored by applying a stream of air to blow away the oil, coolant, or other liquid on the surface. Therefore, the presence of cutting fluids does not pose any particular problems. Also, the frequency response and slew rate of the optical system are limited such that the presence of an occasional chip in the optical path will not cause any undue errors.

The measurement scheme utilized by the optical gauging system involves a low-power modulated laser source focused on the workpiece surface. The machined workpiece surface has a nonspecular property which scatters the light yielding sufficient magnitude and phase information to provide a servo tracking signal which relates the distance from the workpiece to a fixed reference station. In this research program, the optical servo system realized and implemented was the result of a study that involved different optical geometries and detector systems and polarized and nonpolarized laser sources. The performance of the optical system was tested against an extreme range of possible workpiece surface conditions. The last effort involved the consolidation of the sensor with the machine and a subsequent study of the closed-loop response of the resultant system.

A modulated HeNe gas laser was selected for use in the gauging head. A square-wave amplitude modulation scheme was utilized to eliminate drift and effects of ambient light. From elementary modulation theory, the modulation frequency must be double the bandwidth of the signal (in this case, the measurement of the workpiece). A high modulation frequency should be used to place the modulation above the noise frequency caused by the tangential translation of the workpiece. The relevant parameter for wideband noise from the workpiece is the ratio of modulation frequency to the photodetector bandwidth, which should be as high as practical at frequencies where intensity modulation from the moving workpiece is significant. Although a solid-state laser would have been more attractive in terms of size, there were not any readily available continuous-wave solid-state lasers at the time this work was done, and pulsed lasers were ruled out because of increased noise due to aliasing with the modulation at the workpiece surface.

GEOMETRIC PRINCIPLE OF THE OPTICAL GAUGE

The random surface grain of actual machined surfaces causes severe distortion of the measurement process. The dependence of the measurement results on the physical parameters of the optical gauging system and the workpiece surface properties was not predictable from theoretical considerations. Moreover, empirical results did not show an obvious quantitative relationship with the parameters of the optical gauging system. However, some qualitative results were drawn from the experiments and some quantitative relationships were derived for a limited range of parameters.

Various techniques were considered for making the distance measurement. It was concluded early on that some sort of optical technique offered the greatest promise for making precise

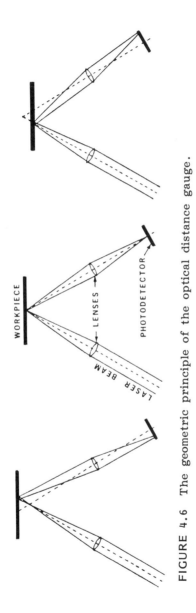

FIGURE 4.6 The geometric principle of the optical distance gauge.

measurements from a remote position. Of several distance-measuring techniques, a geometric approach was taken in which the
laser beam is focused on the workpiece and the resultant image
is analyzed by a position-sensitive photodiode. The distance is
determined from the resultant geometry by triangulation. Interferometric techniques were determined not to be suitable for
this application since they generally require good reflectors and
they give relative and not absolute readings.

The geometric principle is illustrated in Figure 4.6. The
incident laser beam and the centerline of the photodetector are
focused on the same point. The detector lens images the scattered light from the workpiece surface onto the photodetector.
As shown in the figure, the image of the scattered light moves
across the position-sensitive photodetector as the surface of the
workpiece moves away from or toward the focal point of the optics. A linear position photodetector was selected over other
photodetectors because of its simplicity of design and its high
resolution at high incident power levels. Moreover, gaps between diodes in linear array photodetectors blank out part of
the incident power density profile, which, because of the nature of the reflected signal from the surface, may cause measurement errors in this system. This movement across the detector provides the basis for the tracking signal used by the
servo loop, which attempts to maintain the focal point on the
workpiece surface using precision stepping actuators. The distance is determined geometrically relative to the fixed reference
station. The measurement can be achieved either by moving
optical elements within the head or by moving the entire head
on a precision slide. The latter approach was implemented
since there were no problems with moving the head and fixed
optics made the head less complex and physically more compact.

It is possible, over a short range, to make a measurement
without any movement. The accuracy and the range of this
type of measurement are, however, limited by several factors,
including the size of the detector, the accuracy over this size,
and the magnification and depth of field of the optics.

The geometric measurement technique is simple in principle
and the analysis of the measurement on an ideal, uniform diffuse surface is trivial. Indeed, the performance over any math-
ematically definable surface can be determined. However, the
random grain of real machined surfaces causes severe distortion

due to random local variations in reflective properties and multiple reflections and scatter in the surface grain. Theoretical calculation of these effects was found not to be feasible and therefore measurement results had to be determined empirically. This empirical characterization was a monumental task, since results could not be predicted or extrapolated reliably.

The parameters related to the optical gauging system and the orientation of the workpiece with respect to the system can be calculated. What was not readily determinable was the dependence of the measurement results on the surface properties of the workpiece. These had to be determined empirically. Moreover, the results were neither obvious nor predictable. Because of the number of degrees of freedom, complete characterization of the optical principle used herein was beyond the scope of the project. It is recommended that some sort of automated testing machine and data acquisition system be utilized in order to investigate fully this optical principle.

The geometric parameters of the optical gauging principle are analyzed here because they are the basis for the entire optical system. The theoretical geometric parameters and the practical aspects of the measurement system are interrelated. Some of the experimentally observed results, which are discussed later, are considered here since the motivation depends on the experimental results.

The workpiece surface is assumed to be locally planar with changes in reflectivity resulting from nonuniformities in the machined surface. Any translation of the workpiece relative to the optical head can be resolved into a component normal to the surface and a component tangential to the surface. The normal component represents a change in the distance from the head to the workpiece; it is this change to which the head responds. The tangential component merely represents a change in the part of the surface being measured and does not represent a change in distance. A normal view of the plane of the incident beam is shown in Figure 4.7. The measurement plane is defined by the axis of the incident beam and the optical center of the photodetector lens. The line representing the workpiece surface is the intersection of the workpiece surface with the measurement plane, and these planes need not be perpendicular. Line Q is the projection of the line normal to the workpiece onto the measurement plane, and all angular measurements are

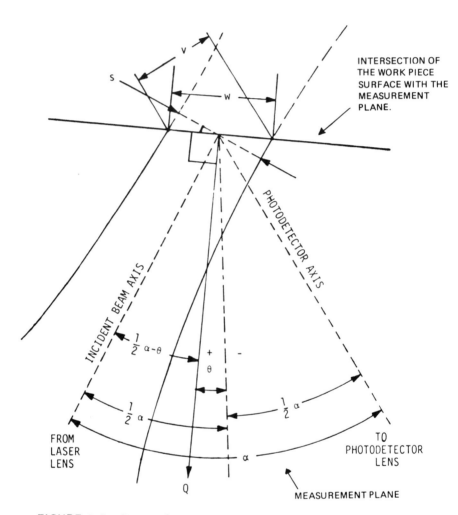

FIGURE 4.7 Geometric parameters and the spot width observed from pertinent directions.

made within this plane. The angle α is the angle between the axis of the incident beam and the axis of the photodetector, which is a fixed characteristic of the typical optical head. The angle of the workpiece surface is referenced to the primary axis of the specular reflection because, although the workpiece is nonspecular, the best results were obtained with this angle. The angle θ is the deviation from this angle.

The incident beam is focused to a spot diameter s at the workpiece surface. Due to nonuniform reflectivity, the apparent position of the incident light varies, resulting in a measurement error. This error can be minimized by reducing the spot diameter. The width w of the workpiece that is illuminated is given by

$$w = \frac{s}{\cos(\alpha/2 - \theta)} \tag{4.1}$$

The width v that the photodetector sees is given by

$$v = w \cos\left(\frac{\alpha}{2} + \theta\right) = s \frac{\cos(\alpha/2 + \theta)}{\cos(\alpha/2 - \theta)} \tag{4.2}$$

These expressions predict that w and v increase as the workpiece is turned toward the photodetector and therefore the error should increase. Experimentally, much better results were obtained with the workpiece normal to the photodetector rather than normal to the incident beam, and the best results were obtained with the workpiece oriented at the angle for specular reflection. Therefore it was felt that the behavior was governed by more complex factors. The spot diameter s is an important parameter, but v and w were found to be irrelevant.

The effect of the normal component of workpiece translation is shown in Figure 4.8. The distance z is the translation of the intersection of the workpiece with the measurement plane. The corresponding translation of the illuminated area observed from the direction of the photodetector is given by

$$x = z \frac{\sin \alpha}{\cos(\alpha/2 - \theta)} \tag{4.3}$$

This equation shows the geometric sensitivity to changes in the workpiece distance observed at the photodetector, but it is not a fundamental limitation in performance because the image of the scattered light from the workpiece can be magnified. This relationship is useful for finding the magnification required to obtain the desired sensitivity, and it also shows the relative effects of displacements of the optical components relative to measured distance.

The true measure of geometric sensitivity is the relationship between z and the tangential distance y on the workpiece. A shift in the apparent position of the incident light due to

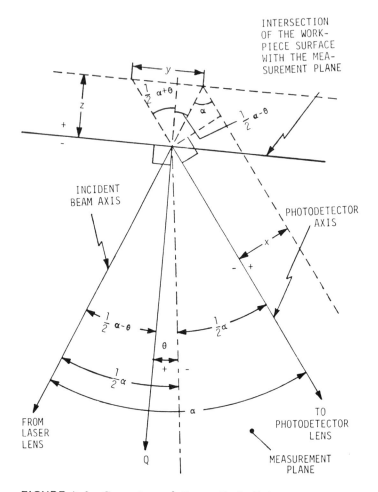

FIGURE 4.8   Geometry of the optical distance gauge.

nonuniform reflectivity causes a corresponding change in z which
is indistinguishable from an actual change in distance. If the
behavior of the light at the workpiece is known, the accuracy
can be calculated for different angles. The relationship between
y and z can be shown to be

$$z = y \, \frac{\cos 2\theta + \cos \alpha}{2 \sin \alpha}$$ (4.4)

The proportionality between y and z is the geometric sensitivity, which increases with α and also increases with θ in either direction away from specular reflection. Experimental results agreed with this relationship for $\alpha \leq 45$ degrees. When α is increased beyond 45 degrees, the accuracy does not continue to improve much, indicating adverse changes in the behavior of the incident light. The accuracy is the same or worse away from the specular direction, which proves that the "specular" reflection angle is optimum even for a nonspecular surface.

An alternative formulation of Eq. 4.4 is useful for analyzing empirical results and predicting or interpolating the results in designing an optical head. The alternative formulation uses an arbitrary empirical factor to compare the experimental results from various configurations of the optical head and find the parameters which give the best accuracy relative to the expected accuracy. With the alternative formulation, the conclusions stated above are obvious when the results are plotted graphically.

Other geometric considerations include the effect of an off-normal surface (not perpendicular to the measurement plane) and a large change in the distance z away from the focal point. A nonnormal surface results in a cosine error in the detected change of position. A large change in distance causes a change in the geometry because the reflected light no longer has the same angular relationship with the detector; however, in practice usually the depth of field of the optics is the limiting factor, so the geometry does not change significantly within this primary limitation.

The concept of referencing the optical system parameters to the plane of the workpiece was useful for developing the geometry. Experimental results from an optical bench were obtained with sample workpieces on an XYZ positioner to permit normal and tangential motions relative to the optical system. However, in a dynamic measurement application, it is not particularly convenient to separate the movement of the workpiece into instantaneous normal and tangential components.

A possible measurement application is shown in Figure 4.9, where the contour of the workpiece is being scanned. The distance and the angle of the workpiece surface change dynamically, and the surface does not move tangentially as the overall workpiece moves. The accuracy in measuring contour varies as the angle varies, and the dynamic position of the point being

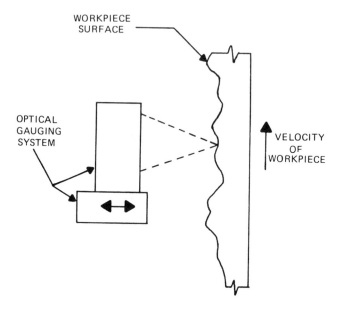

FIGURE 4.9   Scanning a workpiece with the optical gauging system.

measured must be taken into account. This situation becomes rather complex if the locus of points measured by the optical gauging system is not on a line normal to the overall velocity of the workpiece as shown.   An alternative formulation describes these effects.

Many other factors can affect the measurement.   These include the type of photoemitter and detector, the spectral density and cross-sectional density of the incident beam, the incident power, the type of modulation, the diameter and quality of the lenses, and various mechanical factors.   These may change, but the optical measurement process remains the same. For example, the incident beam must have sufficient power to maintain an adequate signal-to-noise ratio, but the optical properties of the measurement are independent of the incident power.

A practical measurement system is shown in Figure 4.10. This optical head has a focal distance of 7.8 in. (20 cm).   The mechanical factors which affect accuracy are discussed here.

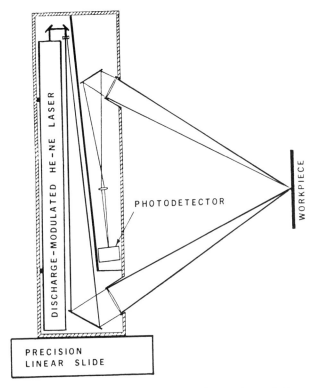

FIGURE 4.10   Consolidated configuration of the optical gauging head.

Thermal expansion affects all mechanical measuring systems, from rulers and micrometers to the most complex gauging fixtures. If the optical head illustrated is fabricated from aluminum, which has a coefficient of expansion of 25 ppm per degree centigrade, expansion will result in a change of 0.0002 in. (5 μm) per degree change in temperature. A corresponding thermal expansion of the fixture holding the workpiece may compensate this to some degree, but many dissimilar materials and coefficients of expansion along with temperature gradients encountered in a typical situation may cause too complicated a computation to quantify. Because of this it is recommended that the temperature be closely controlled or that the measurement be temperature corrected via a calibration curve. Moreover, Invar or other materials with small coefficients of expansion may be utilized.

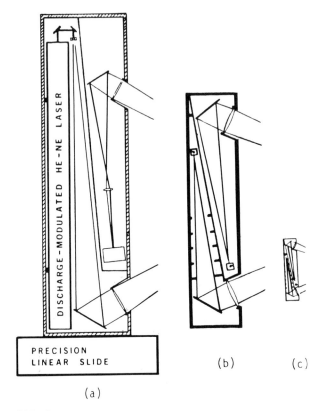

FIGURE 4.11   (a)   Helium-neon laser optical head with a focal
distance of 20 cm.   (b)   Solid-state laser optical head with a
focal distance of 20 cm.   (c)   Solid-state laser optical head
with a focal distance of 5 cm.

All components in the head are subject to mechanical strain
and vibration. These effects are complex and beyond the scope
of this chapter; suffice it to say that this must be taken into
consideration when designing an optical head.

If a complete system analysis of errors is desired, the incremental error for each degree of freedom can be found. The total
error produced by any combination of perturbations can be found
by superposition, which is analogous to the use of partial derivatives, and the results are applicable to small linear changes.
Any method for analyzing incremental changes can be used.

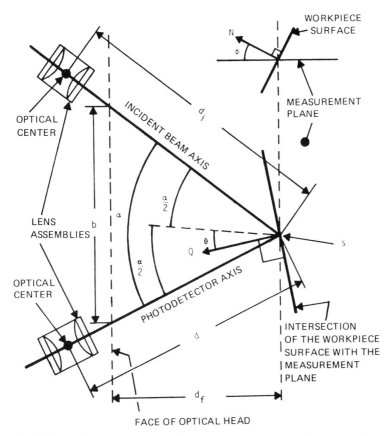

FIGURE 4.12 Summary of the geometric parameters and the physical dimensions of the optical head.

Other techniques must be used for analyzing gross errors or effects that disrupt normal system operation.

The overall accuracy of the system will be limited by the accuracy of the optical head, since the accuracy of available positioning slides is much greater than that of the head. The measurement range, on the other hand, is limited only by the range of the positioner.

The optical gauging principle presented here can be implemented in a variety of configurations to meet physical requirements. The configuration depends primarily on the gauging

distance and the sizes of the optical components used. In this
study a modulated HeNe laser was utilized, and it happened to
be the largest single component. Figure 4.11 shows the configuration compared with configurations using solid-state lasers,
which are considerably smaller. Even with the light paths
folded to minimize size, Figure 4.11a shows the 12.6-in. (32-cm)
laser with a large relative size compared with the solid-state
lasers depicted in b and c. Figure 4.11b shows a significant
reduction in size for the same focal length, while c shows an
even smaller configuration with a reduced focal length.

The geometry and dimensioning of the optical sensor are
summarized in Figure 4.12. The measurement plane is defined
by the axis of the incident beam and the optical center of the
photodetector lens. The geometric parameters which determine
the performance are the focused spot diameter s of the beam
and angles α, θ, and ϕ. The physical dimensions of the optical
path are the focal distance d_f (also known as the standoff distance), the baseline b, and the distances to the optical centers
of the lenses d_i and d. Q is the projection of N in the upper
right of the figure onto the measurement plane.

LIMITATIONS DUE TO MACHINED SURFACES

The single greatest limiting factor on the accuracy of the system was found to be due to the workpiece itself. It limited not
only the accuracy but also the ability to calculate or predict
performance. The ultimate accuracy and dynamic performance
are determined primarily by the surface condition of the workpiece. Because of the random nature of the grain in machined
and other nonspecular surfaces, the performance of the system
could not be determined numerically and had to be found
empirically.

When the geometric parameters of the system were examined,
the workpiece surface was assumed to be planar, with nonuniform
reflectivity to account for the importance of the spot diameter.
Actual workpieces have complex, random, three-dimensional surfaces which are not amenable to mathematical treatment. So-
called flat workpieces are not planar and are not optically specular or ideally diffuse.

Many machining operations, such as turning or grinding,
produce a straight directional grain pattern consisting of ridges

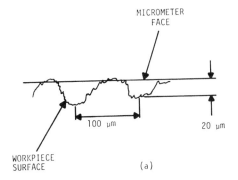

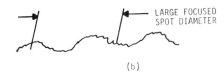

WORKPIECE
SURFACE   (a)

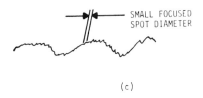

FIGURE 4.13   (a)   Measurement of peaks of the workpiece surface with a micrometer.   (b)   Averaging of the surface contour
with a large focused spot diameter of the incident beam.   (c)
Resolution of the surface contour with a small focused spot diameter of the incident beam.

and valleys.   The angle of the directional surface grain relative
to the intersection of the measurement plane with the workpiece
surface affects the results of the measurement and this angle is
an important factor.   This directionality was not included as a
geometric parameter because it is not a physical characteristic
of the optical head.   Some machining operations produce a fine
random grain pattern, while others, like an end mill, produce
a swirling pattern with locally variable direction.

Fundamentally, there is a difference between a measurement
made with a laser beam or other noncontact device and a measurement made with a micrometer which could cause correlation
differences. The cross section across the grain of a turned
workpiece is illustrated in Figure 4.13. As shown in Figure
4.13a, the micrometer measures the peaks of the ridges over a
wide area. Noncontact methods measure some average of the
surface with the distance averaged over some area as a function
of the spot diameter, as shown in Figure 4.13b and c for two
cases, a large and a small spot diameter. The optical head cannot resolve changes in distance within the illuminated area but
it can resolve contours somewhat larger than this area. Since
this optical head is directional, this averaged measurement is
also a function of the direction of the grain.

The image produced by most light-based systems is a spot
with nonuniform illumination and power density. The image has
an illumination profile which is typically highest in the center
and falls off in some manner, not always monotonic, away from
the center. Most of the incident power is within the diameter
and the illumination or power density is much lower toward the
edge of the spot, continuing over a large area outside the effective spot diameter. This fact, coupled with the unknown surface condition, will also affect this averaged value. To com-
pound this, the beam may not have radial symmetry.

Within the effective spot diameter or width, the unresolva-
ble artifacts of the surface grain cause highly nonuniform reflectivity. The incident beam illuminates the workpiece with a
certain profile. The nonuniformity alters the profile and effectively shifts the apparent position of the beam center. This
effect is illustrated in Figure 4.14. It causes an apparent
change in distance because the measurement process cannot distinguish between a change in distance and the shift in the center of power density. Because of this condition, a small spot
size is necessary for precise measurement. In the experimental
optical head a spot size of approximately 0.0005 in. (12 µm) was
used to achieve 0.0004-in. (10-µm) precision.

In addition to the effects due to the nonuniform reflectivity,
the incident light also scatters and propagates along the surface
by multiple reflections in the surface grain. Therefore, some
random light is scattered from areas which are not directly illuminated by the incident light. This also affects the distance

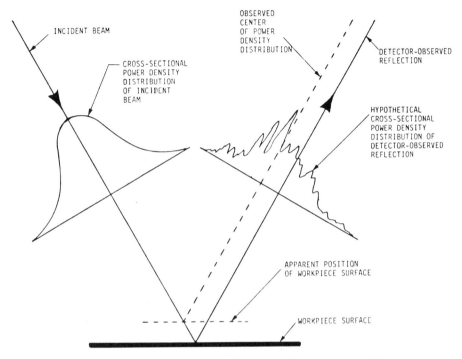

FIGURE 4.14 Optical measurement error due to nonuniform reflectivity of the workpiece surface.

measurement since it causes an effective shift in the beam center.   Polarization of the light will not have an effect since the
surface causes so much distortion of the incident beam.   Even
if the incident light could be focused on an ideal point, the
scattering of the light due to the surface would be the fundamental limitation to the measurement accuracy.

Utilization of an increased spot diameter can lead to an
averaging of the surface, yet this depends on the spot diameter
relative to the spatial frequency distribution of the surface reflectivity including diffractive effects.   The reflective properties
also depend on the wavelength of the incident light.   Lack of
knowledge of the optical properties of machined surfaces limits
the ability to calculate performance.

The scatter and propagation of the incident light along the surface cause some startling and unpredictable effects. The incident light can scatter much farther along the grain than across the grain, and thus the same surface can appear farther away when measured along the grain. In fact, the measured distance can be 0.002 in. (0.05 cm) below the surface of the workpiece along the grain. It may appear that making the incident beam normal to the workpiece could improve results, but actually the results are much worse, even worse than placing the workpiece normal to the photodetector and using a shallow angle of incidence. Much better results are obtained while measuring across the grain, which impedes the scattering of incident light along the surface.

As with the surface effects just discussed, the effect of a liquid film on the surface could not be quantified; accordingly, this effect was empirically determined. The presence of a cutting fluid can affect the measurement significantly.

If the film is transparent, the incident light is scattered from the film and from the surface of the workpiece. In addition, there are multiple internal reflections between the workpiece surface and the undersurface of the film. In the specular direction, the strong surface reflection dominates and the surface appears to be closer. In other directions, the light scattered along the grain causes the surface to appear lower or farther away. Thus the measured surface is shifted one way with specular reflection and the other with other orientations. The multiple internal reflections are more severe along the grain because this is the direction along which the incident light tends to scatter. Translucent films were not studied because of the wide range of characteristics of these fluids. With opaque fluids, obviously a new surface for measurement would be presented to the gauging sensor.

The key point is that an optical measurement system cannot penetrate a liquid film in the way a contacting or pneumatic device can, and it is not immune to the film in the way an eddy current device is. To achieve a good measurement, the film must be blown away with an air source in order to keep the liquid from interfering with the measurement.

THE OPTICAL SERVO LOOP

The optical gauging system itself is a dynamic measurement system with certain response characteristics. When it is used within a feedback loop in a gauging application, it is important to understand the response characteristics, since the results of the statistical averaging depend on the dynamic response of the gauging system.

In some simple applications such as a flatness or runout gauge, where measurement over a limited range is satisfactory, the system can be implemented without moving parts by using a photodetector such as a photodiode array capable of measuring the position of the scattered light image. The response of the photodiode array is more complex, however, than that of the linear position sensor.

In a gauging system without moving parts, the distance is geometrically related to the position of the light on the photodetector, and for small displacements where angular conditions are constant the result is nearly linear. The system described using a linear position sensor is illustrated in Figure 4.15. A digital output is obtained by sampling and converting the signal outputs before or after normalization. A photodiode array can provide a direct, quantized position indication, and there is the possibility of forming an average of the quantized positions weighted by the incident power at each diode location but statistical averaging of a number of direct quantized measurements is superior. The overall range of the measurement is ultimately limited by the depth of field of the focusing optics.

In a system which measures over a wide range, a servo control loop drives the optical system with a precision lead screw to maintain the image of the scattered light centered on the photodetector. The position is then translated by a digital shaft encoder. The response of the system is limited by the mechanical response of the servo loop, but there is a complex relationship between the mechanical servo loop response and the electrical response of the photodetector.

A servo-controlled system based on a stepping motor is illustrated in Figure 4.16. A simple center-null detector is sufficient to provide a digital tracking signal which determines the direction of the servo, but this type of control is not suitable

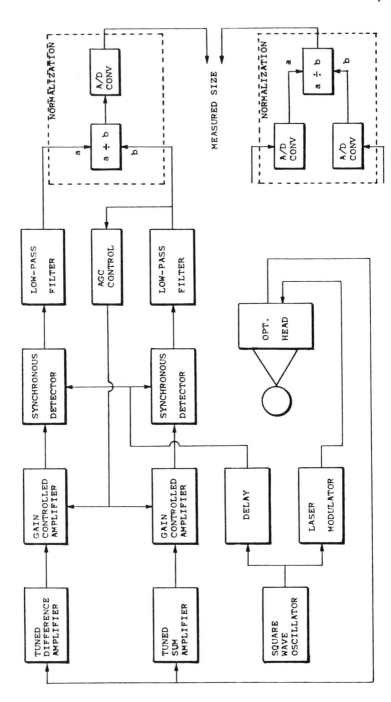

FIGURE 4.15 Optical gauging system using a linear position sensor for measurement.

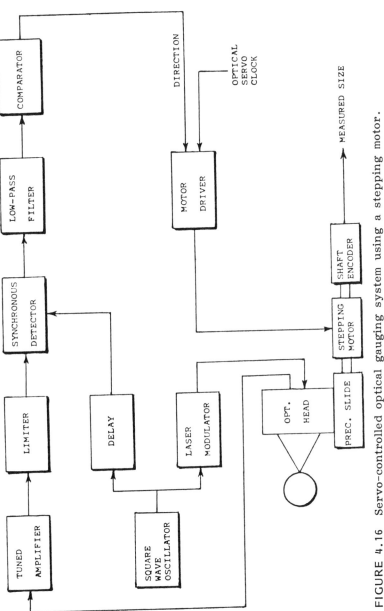

FIGURE 4.16 Servo-controlled optical gauging system using a stepping motor.

for a linear servo control loop, while a stepping motor is ideal
for this type of control. During normal operation of this optical system, the motor steps continuously and the direction of
each step is determined by the output of the photodetector.
The optical system tracks the workpiece at the stepping rate of
the motor with one discrete sample of the measurement produced
for each motor step. The characteristics of this measurement
system are the same as those of a continuous approximation A/D
converter or delta modulator operating at the stepping rate of
the motor.

The delay in the reference for the synchronous detector
(correlator) is used as a phasing adjustment to allow for delay
in the signal path due to electronic components, such as the
delay of the discharge-modulated laser. The presence of the
limiter and the bandwidth of the tuned amplifier and the low-
pass filter affect the response of the photodetector compared
with the response of the linear detector. This has a slight
effect on the overall servo response. Since the limiter mini-
mizes the weighting due to rapid changes in reflectivity and
limits the response to large momentary changes in distance,
the effect of the limiter is generally beneficial.

It is possible to use the digital tracking signal in a linear
servo control loop, but the lack of a linear control region in-
variably causes oscillation of the loop. With proper design the
oscillation can be limited to the desired resolution, but then the
rapid response of the linear servo motor is lost, making the
stepping motor preferable. For proper operation of a linear
control loop, an optical head with distance-measuring capability
over a small range, as shown in Figure 4.15, must be used to
obtain the maximum dynamic response. The resulting system
is more complex because both true position sensing and a servo
control loop are required. Mathematical analysis of the dynamic
response of the linear servo with a linear position sensor in the
presence of additive noise is greatly simplified by the simpler
treatment of linear devices.

The choice between a linear servo and a stepping motor
depends on the application. A linear servo system has rapid
small-signal response and a very high slew rate. Ordinary
stepping motors have rather low random-stepping rates on the
order of 100 Hz. These motors can be accelerated to much
higher slew rates, on the order of 2 kHz, but the lack of a

linear control region does not permit advance deceleration of
the motor to avoid overshoot. With a linear control region,
higher slew rates can be used, but the small-signal response
is still limited to 100 Hz.

In machining applications, where the overall distance changes
slowly but there is a rapid small-signal variation in distance due
to surface grain, a very rapid small-signal response is desirable,
but a high slew rate is irrelevant and a linear response is undesirable. In such applications, the measurement is subject to
large momentary deviations in the measurement due to occasional
chips on the workpiece surface or in the optical path. The system in Figure 4.16 does not respond to such disturbances because the slew rate is limited, and the limiter stage prevents
the large impulsive disturbance from generating a prolonged
error pulse in the low-pass filter. In this application, a wide
bandwidth is desirable in the tuned amplifier before the limiter
to avoid prolonging the duration of an impulsive deviation before
the limiter limits the amplitude. A linear position sensor (without the limiter) which responds to the time-averaged mean position instead of the time-averaged median position is adversely
affected by a large momentary disturbance, and the large resulting error output causes tracking of the impulsive disturbance
for a considerable time before its effect is averaged out. In a
linear servo control system, the effect is even more severe, because the large momentary disturbance also increases the slew
rate of the servo, so the disturbance is followed to an even
greater extent. Therefore, for such applications a limited large-
signal response is desirable. A very rapid small-signal response
or stepping rate is desirable to permit statistical averaging of a
number of independent readings in a short time to reduce random error due to the surface grain. Since the readings are
always correlated by slow servo response, rapid response increases the possible averaging speed.

In position-sensing applications such as automated assembly
and robotic devices, the overall distance may change rapidly
and a very high slew rate may be required, along with rapid
and accurate small-signal response. In such applications a
linear control system must be used. Such a system responds
linearly to large momentary deviations, but fortunately these
disturbances do not normally occur in position sensing. A true
linear gauging system is also advantageous in position sensing,
because it is often used directly as a feedback element, and

nonlinear gauging response characteristics may have adverse
effects on the overall control application.

It is possible to implement hybrid measurement systems to
obtain extremely rapid small-signal response over a large overall range which does not change rapidly. This can be achieved
with the optical gauging system in Figure 4.15 mounted on a
mechanical positioner with a precision lead screw, as in Figure
4.16. Although continuous mechanical servo operation with si-
multaneous rapid electronic small-signal measurement is theoretically possible, the instantaneous ambiguity of the mechanical
position of the optical head due to the dynamic mechanical response characteristics introduces an error in the electronic
small-signal measurement. Such a system is entirely feasible
if the mechanical servo is stopped after initial positioning for
independent electronic measurement. In fact, a wide bandwidth
may be used for the electronic distance measurement illustrated
in Figure 4.15, with a subsequent variable low-pass filter to
provide a flexible measurement system with various combinations
of mechanical and electronic distance measurement capabilities.
Such a system provides a variety of measurement response characteristics which can be tailored to any measurement situation.

Practical considerations for all optical heads are the resolution and the overall range of the photodetector. A split-image
photodetector without gaps inherently has a very high resolution and an arbitrarily large overall range, because the sharp
transition occurs at the center and the overall range away from
the center is physically limited only by the area of the image
splitter and the photodiodes. With a very small image on the
photodetector, position resolutions of better than 0.1% of the
active range for complete transition are feasible, and at high
light levels the median of the incident light can be resolved to
a much finer degree, better than 1 part per million.

Position sensors must have the required resolution over the
required measurement range. The required resolution is determined by the possible gauging accuracy, and in the case of photodiode arrays a compromise is necessary to limit the number of
array elements. The overall range is ultimately limited by the
depth of field of the incident beam, and a smaller range may be
adequate for some applications. The lowest relative resolution
is actually required for the most precise optical heads, because
the small spot diameter has a small depth of field, on the order

of 0.012 in. (300 μm), with an inherent measurement error of
±0.0004 in. (±10 μm). A photodiode array with 30 elements
gives marginal resolution over the maximum range, and 120 elements are more than adequate to achieve the ultimate precision.
With larger spot diameters, the error does not increase propor-
tionally, while the depth of field increases as the square of the
spot diameter, and much greater relative resolution can be
utilized.

For proper operation of a servo-controlled optical system,
the photodetector should be able to detect light relatively far
away from the centerline, so that the gauging system can initially "find" the surface away from the focal point. Although
the incident beam may be severely defocused away from the
focal point, and precise distance measurements are not possible,
the large shift from the photodetector axis ensures a definite
indication of the direction of deviation from the center. The
same photodetector must have relatively very high resolution
near the centerline for accurate measurement. The split-image
photodetector achieves this, but special fabrication is required
if a position sensor is used in conjunction with a servo control
loop. In a gauging system without moving parts, an overrange
is not essential, but it may be desirable to indicate beyond
which end of the measurement range the workpiece is. To
achieve this in a photodiode array without using hundreds of
array elements, the requisite number of elements can be fabricated near the center with a large photodiode area at each end
for a large overrange capability. The linear position sensor is
not limited by discrete elements, but thermal noise may limit
the relative resolution. An overrange capability may be added
by greatly lowering the sheet resistance of the ends of the device compared with the central region, so that most of the output current transition occurs in the central region. If sufficient incident power is available at the photodetector, the advantages of the linear position sensor are apparent, and the superi-
ority of accurately referencing the position to the center null
point is obvious. The center null point is always a null regardless of incident power or normalization, and scaling error due
to normalization is a percentage of the distance away from the
null point. The linear position sensor has no relative resolution
limit other than noise, and the standard device may be used
over any measurement range with a large overrange as long as
there is sufficient light. The measurement accuracy depends
on the stability of the center null point, and the differential

transformer eliminates any significant error in null detection. The temperature of the photodetector can be controlled rather easily to eliminate thermal drift of the null point.

Regardless of the overrange capability, the active range of a high-accuracy optical head may be small, on the order of ±0.04 in. (±1 mm), before the reflected light falls completely off the photodetector. Therefore, an override for the optical servo control loop is needed to position the optical head initially. In a machine control system, the optical gauging system compensates for tool wear and eliminates the need for expensive preset toolholders. However, normal tool wear and simple, inexpensive preset toolholders do not exceed ±0.04 in. (±1 mm) errors, so the machine control system should always be able to position the optical head to "find" the surface. Also, the initial size of the workpiece is usually known within ±0.04 in. (±1 mm), so the optical head can be initially positioned regardless of the tool position.

An override for the optical servo control is also needed to prevent the optical system from randomly deviating from the proper range at the end of a workpiece. In case of a large step in the workpiece, the override can cause the optical servo to slew at the maximum rate to the proper new range.

In general, an override for the basic optical servo control loop is required whenever the measured surface ends or changes in a sharp step. The numerical control system with which this optical system is used is suitably programmed to keep track of the dimensions of the workpiece as it is machined, and the optical system is an integral part of the overall system, with the optical servo very likely driven by the numerical control system. Proper control of the optical system is merely a matter of suitable programming, just as successful machining is a matter of proper programming, without which severe destruction can result.

The status of the measurement system can be tested by auxiliary measurement of the photodetector output. The presence of a surface near the focal point can be detected by measuring the sum or total photodetector signal current. If this is otherwise not required, a simple, inaccurate measurement is sufficient. The presence of a detectable signal above a threshold indicates that reflected light is incident on the photodetector. From that point the optical servo can operate normally.

Arrival at the null point is indicated by several direction reversals, which do not occur until the null point is reached or when there is no signal at all. From these indications, the numerical controller can scan or slew the optical system until it finds an arbitrary surface, and the loss of a surface due to scanning past the edge can also be detected.

The basic operation of the optical gauging system is qualitatively simple, but the detailed analysis and calculation of performance is generally very complex and difficult. The use of the resultant measurement from the optical gauging system for numerical machine tool control is relatively simple and intuitively obvious, and the application of the optical gauging system to machine control is a straightforward extension of existing numerical machine tool control technology.

APPLICATION OF THE OPTICAL GAUGE
TO MACHINE TOOL CONTROL

One common feature of all machining applications, where the dimensions of a workpiece are continually changed in process, is that measurement at the point of material removal is not possible, and consequently there is some inevitable delay between the cutting operation and the measurement of the resulting dimension. Because of this delay, a conventional continuous feedback control loop cannot be used for optimum, rapid response. This precludes the possibility of compensating for free travel or backlash in the tool slide lead screw by feedback, and a preloaded lead screw, such as a preloaded ball screw, must be used to prevent sudden changes in the controlled dimension or excessive delay in compensation due to free travel or backlash. A modification of standard servo control techniques permits optimum, rapid response with complete stability in machine control.

As an example, the machine tool control loop is examined for an engine lathe in machining a cylinder. The same control principle can be applied to a variety of machining applications. In other applications where the dimensions of a workpiece are not changed in process, there is no delay in the measurement, and a conventional feedback loop can be used with the dynamic response of the optical gauging system taken into account. The machining of contours is much more complex, and this topic is discussed later in the chapter.

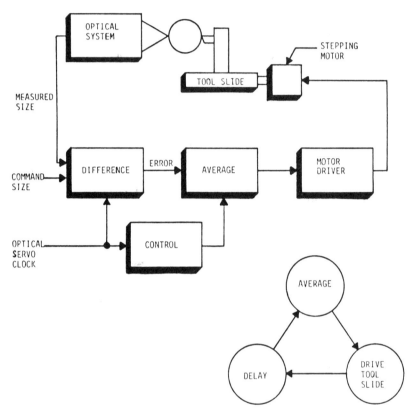

FIGURE 4.17 Basic machine control loop using the optical
gauging system.

A basic machine control loop for machining cylinders on an
engine lathe is illustrated in Figure 4.17. Assuming a digital
numerical control system, the measured workpiece size from the
optical system is sampled and converted in some manner to digital form compatible with the numerical controller, and the optical
servo clock corresponds with the samples of the measured size.
The measured size is compared with the command size, and the
algebraic difference generates an error count, as in a conventional control system. This error count is derived from the
measured size of the workpiece instead of the position of the
tool slide lead screw, and there is some delay between a corrective movement of the cutting tool and the measurement of
the resultant change in size.

As discussed previously, the measured size is noisy, and the random variation in the measurement can be reduced by averaging statistically a number of independent or at least fairly uncorrelated samples of the measured size. For a constant command size, the same mathematical result is achieved whether the measurement is averaged before comparing with the command size or after, but the error count (after comparison) is much smaller in magnitude and number of significant digits than the actual measured size, so numerical computation may be simplified. A more important reason for averaging the error count instead of the actual measured size occurs when the command size changes continuously for machining contours. In this case, averaging of the error automatically accomplishes curve fitting to the command size. This topic is discussed later.

A number of samples of the error are statistically averaged to reduce the random measurement noise and prevent random machining corrections which track the measurement noise. The averaged error is used as a correction for machining.

The tool slide lead screw is driven by a stepping motor. Since a preloaded lead screw, such as a preloaded ball screw, must be used for the tool slide, the stepping motor gives incremental control of the tool slide, which permits positive, predetermined positioning of the cutting tool in fine steps. In a conventional feedback system, the delay in the feedback loop would cause poor servo response. In this system the incremental control of the tool position permits the cutting tool to be moved by an increment equal to the difference between the command size and the measured size, which permits immediate compensation of machining errors after one delay period.

Immediately after the correction in the tool position, the optical system still measures the previous size of the workpiece. It is necessary to maintain the new position of the cutting tool for a delay period until the new workpiece size reaches the point of measurement. After this delay time, the new workpiece size can be measured and averaged for the next correction of the tool slide position.

A state diagram or flowchart of this simple, steady-state machine tool control loop is included in Figure 4.17. In a practical system, a much more complex control system is required to initiate and terminate a cut, override the optical gauging system beyond the ends of the measured surface, and determine whether

the optical measurement is valid according to the status controls
discussed earlier. This control system obviously gives the fastest possible corrective response to machining errors. The accuracy and resolution are limited by the accuracy of the optical
gauging system in measuring the controlled dimension, which is
maintained within the measurement capability (accuracy) of the
optical gauging system.

   In an idealized implementation of this system, the increment
for the tool slide correction should be rounded off to the nearest number of steps of the drive motor so that the best possible
correction is made. In a practical system, it is better to trun-
cate the fractional part of a step to prevent needless corrections
of the tool position. If the accuracy of the optical measurement
is great enough to permit gauging within a fraction of a step of
the cutting tool position, truncation of the fractional part of a
step may give a machining error greater than half of a step,
but it prevents oscillation of the cutting tool by one step back
and forth. If the error is almost one step, it is likely that
some noise will cause the correction to the next step, and then
the new position will be maintained. Technically, an oscillatory
error of ±1/2 step of the tool position is more accurate than a
slightly larger constant error, but there is no practical difference, and a nicer finish is produced when the oscillatory corrections are eliminated.

   The undesirable oscillatory corrections of the tool position
may be further limited by truncating the first few steps of the
correction, and this number can be programmed according to
the required accuracy. For example, if ±0.002 in. (±50 μm)
accuracy is required, a smooth finish which differs by 0.002
in. (50 μm) from the command size is preferable to a surface
with random steps within, say, ±0.0006 in. (±15 μm) of the
command size, even though the latter surface is more accurate.
If the surface quality of this particular workpiece permits an
optical gauging accuracy of ±0.0008 in. (±20 μm), the first
0.0012 in. (30 μm) of the correction increment can be truncated
to achieve the required accuracy, and most correction increments
are eliminated. Even better results are achieved if the truncation is initially reduced to, say, 0.0004 in. (10 μm) to reach the
nominal size and then raised to 0.0012 in. (30 μm) to provide
corrections only when the workpiece is out of tolerance. Of
course, if the surface quality is reduced and the optical gauging accuracy becomes ±0.0012 in. (±30 μm), then only 0.0008 in.

(20 μm) can be truncated from the correction, and some corrections may occur, depending on the amount of random variation in the measured size.

The reason that this type of incremental control can be successfully employed in machining applications is that machining conditions are usually stable and machining errors do not change greatly in a short time. The primary causes of error are wear or improper positioning of the cutting tool, deflection of the tool under varying loads, lead errors in the tool slide lead screw, and long-term changes such as thermal expansion. This system cannot compensate for rapidly varying errors such as chatter or backlash in components, and all other errors are mainly long-term errors. An incremental movement of the cutting tool results in a corresponding similar incremental change in workpiece size.

Incremental control of the tool position is a key element in this control system. The prime mover for the tool slide can be a simple stepping motor or a linear servo motor with independent feedback from the motor, but in either case the ability to move the cutting tool by a known increment is essential. In a conventional feedback system, no definite relationship is assumed between the controlled variable and the drive signal, and a delay in the feedback loop leads to very slow servo response and/or oscillation.

Lead errors in the tool slide lead screw are not important in this system, as long as the tool motion is smooth. The incremental positioning is used to correct the error between the command size and the measured size, which is a very small increment, certainly less than 0.004 in. (100 μm). Any lead screw has negligible lead error over such a small increment, no matter how worn or inaccurate it is. Problems occur only if the lead screw is damaged, for example, if a ball screw has dents in it from severe abuse, or there are chips in the ball nut. A dirty or damaged lead screw has an irregular motion which causes unpredictable positioning, and proper operation of the control loop is disrupted.

In machining extremely tough materials, the relationship between the cutting tool position and the resulting workpiece size does not track on a 1:1 basis. There may also be some lead error in the tool slide lead screw or other errors in the

control system. To avoid overshoot in such cases, the correction increment for the tool slide may be deliberately multiplied
by a positive constant less than one, such as 90%. This constant can be changed to suit the machining conditions and desired results. The use of this constant causes the workpiece
size to approach the command size exponentially in a stepwise
manner for each cycle of the correction procedure, and the
margin in each correction increment prevents overshoot. The
multiplicative constant may be used in conjunction with truncation of the correction increment, described previously.

The complete operation of this control system is intuitively
obvious, and the quantitative results can be calculated incrementally by inspection under any conditions, with the results
being as accurate as the ability to model the machine and the
machining operation. Due to the incremental, cyclic control of
the machine, analytical mathematical treatment is neither necessary nor reasonable.

The relevant parameters in this machine control loop are the
delay time and the averaging time. The delay time depends on
the machining conditions, and it will be discussed subsequently.
The averaging time depends on the desired number of samples
and the correlation time of the measurement. The correlation
of the measurement is due to the response of the optical gauging system and the surface speed of the workpiece. A mechanical servo in the optical system can operate at 1 kHz, which
gives 16 fairly uncorrelated samples in 16 ms. The spatial correlation of the surface grain may have a correlation length on
the order of 0.002 in. (50 μm), so several hundred micrometers
of the surface must be scanned for a good statistical average.
At extremely low surface speeds, such as occur in milling, a
longer averaging time may be needed in order to measure a sufficient segment of the surface, but cutting conditions do not
change significantly in such a short interval.

An important practical consideration in this basic control
loop is the time for the optical gauging system to respond to
the abrupt correction in workpiece size. The delay time is the
time for the machining correction to reach the measurement
point. The time for the mechanical servo in the optical gauging system to step to the new size is usually negligible for a
small correction, and this time can be incorporated in a small
margin in the delay time, but for a fairly large machining

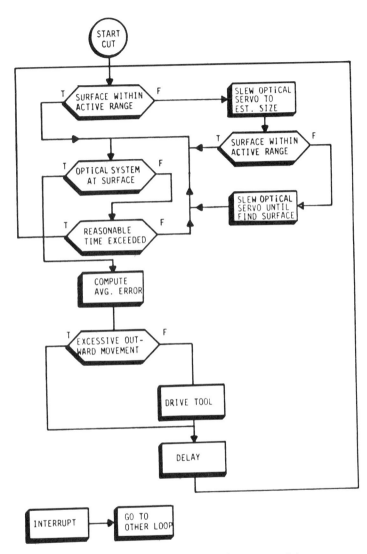

FIGURE 4.18 More complete machine control loop.

correction, the response time of the optical gauging system can be significant compared with the delay time. If a short delay time is important and the additional time for the optical system to respond to a large correction is objectionable, an additional step can be inserted between the delay operation and the average operation to test the status of the optical gauging system and wait only until the optical system reaches the surface. The override of the optical servo control loop can also be used to preposition the gauging system at the anticipated new workpiece size during the delay time. Thus the overall servo response time is minimized.

A complete practical numerical machine tool control system is very sophisticated, and a complete representation is beyond the scope of this chapter. Some practical features for a control system are presented here.

A flowchart for the basic operation of a machine tool control system is shown in Figure 4.18. Some of the steps in the flowchart comprise many operations, but the operations are obvious and need no detailed diagrams.

At the first cut, the cutting tool may approach the workpiece from a random initial position. The machine must be positioned so that the optical system measures the workpiece, and the cutting tool should preferably be at or away from the workpiece surface to avoid damage. When the cut is started in Figure 4.18, it is first necessary to determine whether the workpiece is within the active range of the optical gauging system. The presence of strong reflected laser light may be tested at the photodetector, or other criteria may be used. It may be assumed that the actual surface is near the command size, because some material is to be removed, but an extremely heavy cut causes damage and must be avoided. Therefore, the optical system may be assumed to be within the active range if it is within a certain range about the command size. Alternatively, the approximate size of the workpiece may be known, so the optical head can be positioned appropriately. In any case, if the surface is within the active range, according to a suitable criterion, the operation can proceed. Otherwise, the optical system must be positioned by the controller to the proper range. If predetermined positioning fails, the controller must slew the optical head over the entire travel until the surface is found.

After the optical head is positioned so that the surface is within the active range, the optical servo (which may be part of the same machine control system) must operate normally until the actual surface is reached, which is indicated by direction reversals at the focal point. If this is not achieved within a certain time limit, the active range of the optical system may be retested.

When the optical gauging system reaches the surface, so that a valid measurement is obtained, the machining sequence can begin. The tool approaches the workpiece in increments equal to the difference between the measured size and the command size, or a modification of this, as discussed previously. There is a delay after each movement of the cutting tool to permit the optical system to measure the resulting size.

The tool feed should start on or before the moment the tool contacts the workpiece in order to maintain the proper relationship between the tool and the gauging system and ensure proper transit of the workpiece surface from the point of cutting to the point of measurement. After the tool starts cutting, the optical system measures the size being produced and corrects to the command size in the next increment.

After each delay period, the control system tests the validity of the measured size before starting another average and machining correction. Under normal operation, the optical system should always be within measurement range of the surface, but some time may be needed to respond to a large machining correction. This machining sequence continues until it is interrupted.

Normally, the command size is smaller than the measured size, so the tool approaches the workpiece to remove material. Under some conditions, the workpiece may be smaller than the command size, and the cutting tool moves away from the workpiece. Since it is impossible to replace material and increase the size of the workpiece, the tool can move a considerable distance away from the workpiece. If a point is reached where the workpiece is slightly larger than the command size, considerable time is required to return the tool in small increments to the workpiece. Therefore, a test for excessive outward movement of the cutting tool is included in the loop. Every time an average is computed, the resulting correction is checked, and if the total outward movement of the tool is excessive, the movement of the tool is bypassed. The criterion for this test depends on circumstances. A useful criterion assumes that a

few consecutive outward corrections are normal regardless of
the magnitude, and a considerable number of consecutive outward corrections is possible, but this total travel is small.
For example, after three consecutive outward corrections, the
total additional outward travel is accumulated, and when it
reaches 0.0016 in. (40 μm), further outward travel is inhibited.
Any inward travel resets the count of the initial three outward
corrections, but after the three outward corrections, when the
total outward travel is accumulated, inward travel only decrements the total outward travel by the inward travel. The purpose of this is to prevent long sequences of outward travel
reset by occasional inward travel when the actual size is just
under the command size. After the inward travel decrements
the accumulated outward travel to zero, one additional inward
step resets the entire count.

When the tool approaches the workpiece from a random initial position and the difference between the command size and
the actual size is small, a considerable time is required for the
tool to reach the surface. There is no indirect way of knowing
when the tool contacts the workpiece, so the feed must run
long before the tool starts cutting. If the tool is electrically
insulated from the rest of the machine, the moment when the
tool contacts the workpiece can be detected by simple electrical
contact with the grounded workpiece. If this is done, the tool
can approach the workpiece continuously at the random stepping
rate of the motor until contact is established, and some machining time can be saved. When the tool contacts the workpiece,
the feed is started and normal machining begins.

In any case, when the tool approaches the workpiece from
a random initial position, it leaves a ridge at the beginning of
the cut. This can be removed by returning the tool to the
beginning of the cut after a known size is established. After
the rough cut, several finishing cuts are usually made. In this
case, the initial ridge is eliminated because the position of the
cutting tool is known after the previous cut. The cutting tool
is moved by the difference between the previous command size
and the new command size to start the new cut properly.

The control system for starting and ending cuts is illustrated
in Figure 4.19. The transition from the loop in Figure 4.18 to
Figure 4.19 is illustrated by an interrupt, although in an actual
numerical machine control system the real-time program can

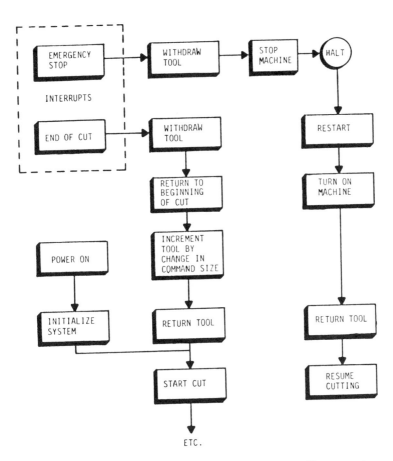

FIGURE 4.19 Control loop for starting and ending a cut.

perform all the control functions and the transfers between loops by normal program execution of in-line code.

At the end of the cut, the cutting tool is withdrawn from the workpiece by a known number of steps, and the tool can be returned to the same position at the beginning of the next cut to resume the same workpiece size.

When the command size is changed between cuts, the tool slide is incremented by the net change in command size. When the tool is returned to the workpiece, it starts the new cut at approximately the correct workpiece size, and there are no more ridges in subsequent cuts. This incremental change in the tool position by the net change in command size is a very important operation during machining, which will be discussed subsequently. After the cut is started and the optical gauging system measures the resulting workpiece size, normal tool control resumes.

In the basic control system in Figure 4.18, the command size is assumed to be constant. If the command size varies, the machine tool control loop compensates for the error and follows the command size in a stepwise manner due to the delay time. If the cutting tool is always directly moved by increments equal to the changes in command size, as in Figure 4.19, while simultaneously operating in the control loop in Figure 4.18, this system is almost capable of machining smooth contours. If these two motor drive commands are added or linearly superimposed and the command size continuously follows a contour in the smallest steps, the cutting tool follows the same contour, but the optical gauging system does not make the proper corrections in this simple control system. The proper way to do this is discussed later.

The numerical machine tool control system described here is not much different from conventional numerical control. Modern numerical controllers which use general-purpose programmable computers can be reprogrammed to employ the new optical gauging system in the feedback loop, and they probably have sufficient computational capacity to operate the optical servo control loop directly. Of course, in an actual machine, there are many controlled axes, and the overall system is much more complicated than the simple single-axis system described here. Machining of contours requires considerably more data processing, and a more powerful computational system may be required, although an existing system may have enough reserve capacity.

In all machining applications, it is not possible to measure at the point of material removal. In this example, an engine lathe is considered, and similar considerations in other applications are apparent.

A typical machining sequence is illustrated in Figure 4.20. For convenience in representation, the tip of the cutting tool and the point of measurement of the optical gauge are shown at

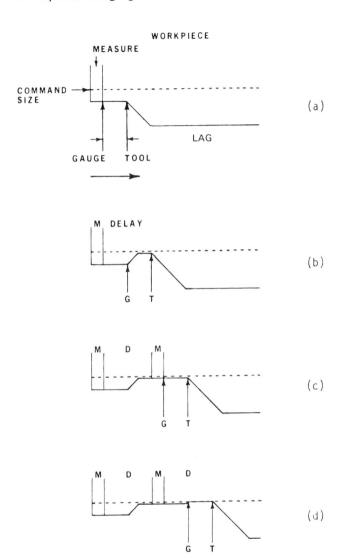

FIGURE 4.20 Typical machining sequence showing the delay in measurement with the tool and the gauge shown on the same side of the workpiece for convenience.

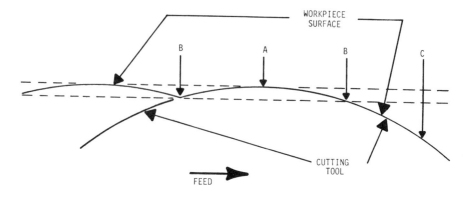

A - LOWEST POINT
B - HIGHEST POINT
C - MATERIAL ALREADY REMOVED HERE

FIGURE 4.21   Exaggerated view of a finish cut.

their respective axial positions on the same side of the workpiece. It is understood that on an engine lathe the optical
gauge measures the contour on the opposite side of the workpiece. The details of the contour due to the feed rate and the
skew of the helical cut are not shown. The time relationship
of the rotation of the workpiece is also not illustrated here,
since the angular position of the optical gauge about the axis
of the workpiece is altered. The optical gauge tracks, for example, 0.002 in. (50 μm) behind the cutting tool. In Figure
4.20a, the optical system measures over the interval indicated
to obtain an average measurement, and the tool starts to make
a correction. In Figure 4.20b, the tool is machining at the
corrected radius, but the optical system is still measuring the
previous radius. A sufficient delay is used to ensure that
the optical system is measuring the new radius, and then an
average measurement is obtained, as in Figure 4.20c. After
the measurement, the cutting tool compensates for the residual
error, and another delay is initiated as in Figure 4.20d.

In actual machining applications, extreme speed is not
needed in the control loop. In normal machining, the cutting
conditions do not change significantly for several millimeters of
feed, so after the correct size is established, frequent corrections are not needed. The only time the small lag illustrated

in Figure 4.20 is necessary is in starting a cut from a random
initial tool position in very tough alloys which work-harden
from machining, so that it is not possible to cut less than about
0.01 in. (250 μm) at a time. If the end cannot be chamfered to
remove the initial ridge, very fast initial compensation is needed
to minimize the ridge, since it cannot be machined off later after
the correct size is established. This problem is normally minimal,
so a larger delay in the feedback loop does not have adverse effects in most practical applications. An excessive delay time
allows more margin in the control system, while insufficient delay causes severe oscillation.

Another consideration is that the turning tool is not a narrow, tapered point. A typical turning tool has a tip radius of
at least 0.004 in. (100 μm), and a tip radius of 0.04 in. (1 mm)
is frequently used in finishing tools. An exaggerated view of
a finish cut is illustrated in Figure 4.21. The tip of the tool
and the minimum radius is at A, and the maximum radius occurs
at B. However, at C the radius is already similar to A and B,
and withdrawal of the tool cannot replace material at C. A corresponding view of a finish cut drawn to scale using a 0.004-in.
(100-μm) feed rate and a 0.02-in. (500-μm) tool radius is illustrated in Figure 4.22. There is little difference between the
radii at A and B, and a smooth finish is produced because of
the large radius of the tool. Due to the effect at C, the resolution of the machining operation itself is physically limited. A
considerable delay in the response of the workpiece radius to
changes in tool position is inherent in this situation, and a long
delay time is required for stability.

In the experimental system, excellent results were obtained
even with a long delay time. It is much more important to average to a large number of measurements for each correction to
obtain a smooth finish than to make frequent noisy, unstable
corrections.

The optical gauging system designed for measurements at a
distance of 7.9 in. (20 cm) from the workpiece was tested on a
small, light-duty 9-in. (23-cm) lathe. The basic measurement
accuracy of 0.0004 in. (10 μm) could be improved to 0.0001 in.
(2.5 μm) by statistical averaging, while machine deflections were
on the order of 0.0004 in. (10 μm), so the test data merely show
that the optical system performed better than the stability of
the mounting. This lathe is rather small for the optical system

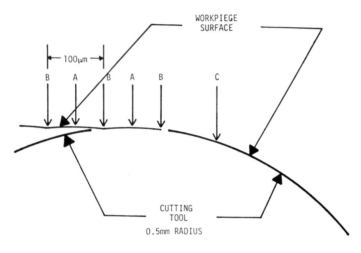

A - LOWEST POINT
B - HIGHEST POINT
C - MATERIAL ALREADY REMOVED HERE

FIGURE 4.22   Actual scaled view of a finish cut using a 100-μm feed rate and 500-μm tool radius.

mounted on the carriage at such a great distance from the machining area.

The optical gauging system can be adapted to a variety of machine control applications.   All the basic features of machine control applications were discussed previously, and these principles can be adapted to other machines.

This optical gauging system can also be applied to a variety of position-sensing applications.   If the position of the measured object can be incrementally controlled and the optical gauging system is used to determine errors due to misalignment, improper relative positioning of an object, or cumulative errors, such as in automated assembly operations, the same incremental control system can be used to ensure complete stability of the feedback loop.   The delay in the feedback loop is not needed, because changes in the controlled dimension are sensed immediately, but the incremental control isolates the dynamic characteristics of the optical gauging system from the feedback loop.

In many position-sensing applications there is no definite
relationship between the controlled variable and the drive signal, and the optical gauging system must be used as a dynamic
transducer in a conventional feedback loop. In such applications, the dynamic response requirements are much more stringent, and very rapid response is often necessary.

The response of the linear position sensor used without
moving parts or with a linear servo system is readily determined.
However, when a stepping motor is used in the optical gauging
system, the hardware is greatly simplified compared with a linear servo and good performance is readily achievable, but analytical treatment of the system as a feedback element is very
difficult. The most practical method of analysis is to assume
quasi-linear response with a large-signal slew rate limitation
(due to the stepping rate of the motor), as explained earlier.
Since the resolution of the stepping motor in the optical gauging system must be high enough to achieve the desired accuracy, the quasi-linear approximation is practically as good as
the true quantized model. Under these conditions, the slew
rate limitation mainly determines the response of the optical
gauging system, and such a response characteristic can be
used for servo control design.

If the proper simplification of the optical gauging system is
made, the calculated response characteristics can be used for
servo control design. If overall system frequency response limitations permit, some averaging of the measurement from the optical system may be desirable to reduce the effects of random
noise. Even the optical gauging system using a stepping motor
can be approximately treated by analytical means and applied
in servo feedback loops.

MACHINING OF CONTOURS

The control system discussed earlier is suitable for machining
workpieces where the command input does not change significantly during the delay period. A modification of the machine
control system is required for optimum machining of contours,
where the command input changes continuously during machining.

The use of the measurement from the optical gauging system
for machining contours is a relatively simple and qualitatively
obvious extension of the control system discussed previously.

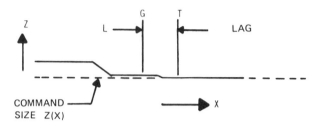

FIGURE 4.23 Relationship between the cutting tool and the optical gauging system with the tool and the gauge shown on the same side of the workpiece for convenience.

Rigorous analysis of machining of contours in actual machining applications is rather involved and beyond the scope of this chapter, but such analysis is a straightforward extension of existing numerical machine tool control technology. A general qualitative description of the considerations and additions for machining contours is presented here.

The machine control loop for machining with a constant command input was discussed earlier, and the salient features applicable to machining of contours were emphasized. A rather minor modification of the control loop permits proper machining of contours. As before, an engine lathe is used as an example of the application of the control loop, but the technique can be applied to other machines.

For initial consideration of machining of contours, the cutting tool and the optical measurement are represented as points which can follow any function of the tool feed, and the tool feed is ignored. For convenience in representing the important relationship between the cutting tool and the optical gauge, these components, shown in their respective positions on the same sides of the workpiece in Figure 4.20, are shown on to the same side of the workpiece at their true position along the axis of the workpiece, as illustrated in Figure 4.23. The distance L is the positional delay or lag between the cutting tool and the optical gauge, and this distance is the feed corresponding to one half of a revolution or more. The distance L is related to the number of revolutions of the workpiece by the rate of feed, and the delay time depends on the angular speed of rotation of the workpiece.

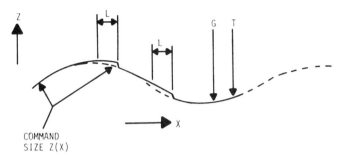

FIGURE 4.24 Correction of machining errors along a contour.

In machining contours, the command size Z is a function of the position X along the axis of the workpiece, defined as $Z(X)$. To achieve the desired contour, the cutting tool must follow the contour at its respective position X_T, so it must be positioned as $Z(X_T)$. The optical system measures the contour at its respective position X_O, so the measured size must be compared with $Z(X_O)$. The contours are related to L, giving $Z(X_O) = Z(X_T - L)$ or $Z(X_T) = Z(X_O + L)$. The values X_T and X_O are functions of time, depending on the actual rate of feed.

For a cylinder, $Z(X_O) = Z(X_T) = $ constant, the system discussed before operates properly. If $Z(X)$ is not constant, the control system must use the proper values of $Z(X)$ for positioning the tool and measuring the machining error.

For proper system operation, it is assumed that the command input $Z(X)$ is defined for the workpiece and any point of $Z(X)$ is available for the controller, as needed. The cutting tool must be continuously driven according to the desired contour $Z(X_T)$, as in a conventional numerical control system, and the corrections from the optical measurement must be added or linearly superimposed to generate the correct workpiece size, as described before. For the proper correction, the optical measurement must be compared with $Z(X_O)$, which obviously generates the desired contour. If there is no machining error, the machined contour X_T subsequently gives no measured error at X_O, and machining proceeds normally. If any machining error causes the machined contour at X_T to deviate from the desired contour $Z(X_T)$, this error is subsequently measured by the optical system; the cutting tool is moved by an increment equal to the

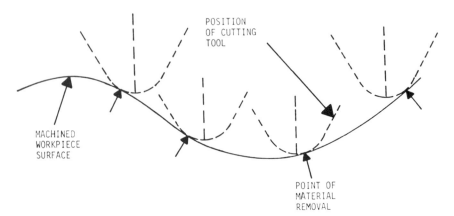

FIGURE 4.25   Variation of the point of material removal relative to the tool tip as the slope of the machined workpiece surface changes.

error, in addition to normal tracking of the contour; and the machined contour is restored to the desired contour $Z(X_T)$, as illustrated in Figure 4.24.   This mode of operation, with the difference between the measured size and the command size $Z(X_O)$ computed for each measurement, inherently permits statistical averaging of a number of readings for any contour.

Proper machining of contours does not depend on the choice of L, as long as there is a sufficient delay to prevent oscillation.   However, the actual value of L must be known in order to maintain the proper relationship between $Z(X_O)$ and $Z(X_T)$. An error in the value of L causes a machining error when the slope of $Z(X_T)$ changes, and the error depends on the second derivative of $Z(X)$ in the region of X_T and the point of material removal.   Of course, a large value of L increases the time required for correction of machining errors and degrades overall results.   When expensive preset tool mounts are not used, the lateral position of the cutting tool and the resulting distance L are not accurately known, just as the radial distance is not known.   Therefore, the effect of errors in L is important.

In actual machining, the cutting tool is not an ideal point. The tip of the tool is rounded, and the sides of the tool away from the tip place limits on the slope which can be machined.

The rounded tip of the tool effectively changes the point of material removal relative to the tool, as illustrated in Figure 4.25, and this affects L. Of course, the tip radius also places a limit on the minimum concave radius of curvature in the contour.

The effect of the shape of the cutting tool is already solved in conventional numerical control systems. By using adaptive control with the optical gauging system, it is possible to compensate for changes in the shape of the cutting tool due to wear.

The extension of existing numerical control technology for effective utilization of the optical gauging system in numerical machine tool control systems is an area for further research.

PRACTICAL CONSIDERATIONS IN IMPLEMENTING THE OPTICAL GAUGE

The theoretical parameters and design considerations were discussed previously. In addition to the theoretical considerations, there are some qualitative practical aspects of successful implementation of the optical gauging system.

The importance of mechanical rigidity and integrity of the optical system was discussed, and a method of analysis of the measurement error due to any perturbation of the optical system was presented. The importance of rigid mechanical design is reemphasized here. The accuracy of the optical gauging system is only as good as the stability of the mechanical components under vibration encountered in typical applications.

The optical components within the optical head require certain mechanical adjustments for alignment and focusing. These adjustments must achieve the required resolution in positioning, and then they must maintain the final settings for a long time under conditions of shock and vibration. Ideally, these adjustments should remain fixed and stable indefinitely, so that periodic readjustments are not needed. The general-purpose optical positioners used for optical bench experiments are usually inadequate; the standard units are usually bulkier than required for a particular application, and the locking mechanisms are not designed for long-term stability under vibration.

For a compact, stable optical head, custom mounts should be designed specifically for each optical component. The

adjustments should be locked by direct metal-to-metal application of stress at all points where movement of the adjustable
assembly can occur, by means of lockscrews, locknuts, or
clamps. If a preloading spring is used, it must supply sufficient force to maintain positive contact without play, and the
adjusting screw must have a locknut to prevent gradual changes
in adjustment. Many adjustments, such as beam-positioning
components, are not extremely critical in practical applications,
where the beam can be visually positioned and aligned along
the approximate desired path. Such adjustments can be made
by the simple means of using slightly oversized mounting holes
on the adjustable assembly. The height of components is best
adjusted with the greatest stability by means of shims instead
of adjustable mechanisms.

     Glass components cannot withstand the required mounting
stress from metal holding devices. If the glass component must
be clamped directly, a suitable resilient cushion should be used
to distribute the stress evenly over the mounting area. With
many glass components, excellent results are achieved by cement-
ing the ground, nonoptical surfaces to well-fitted metal holders.

     Thermal expansion causes measurement errors, but the tem-
perature of critical components can be controlled by various
means. Errors due to weak mechanical design cannot be compensated.

     When the optical gauging system is mounted on a machine,
the reference station or mounting platform is generally not accurately located relative to the measured surface. To achieve
the desired correspondence between the actual dimension and
the measured dimension, it is necessary initially to calibrate
the gauging system at one point. From that point, the measured distance tracks absolutely, so if the measured distance
is correct at one point, it is correct over the entire range.
Of course, the optical system must be physically mounted so
that the actual measurement axis coincides with the required
measurement axis.

     The actual dimension to be measured must be determined
by some other means, such as a gauging block, micrometer, or
other reference standard, and then the range of the optical
gauging system is adjusted to match the actual dimension.
This adjustment may be accomplished by physically moving the

mounting of the optical system, by rotating the shaft and/or the housing of the digital shaft encoder relative to the precision lead screw, or by adding a constant to the measured value. After the initial adjustment is made, the optical measurement is accurate and stable within the stability of the overall system.

Focusing of the incident beam is critical and difficult to observe directly due to the small spot diameter. Focusing of the photodetector lens is less critical, but also more difficult. If a monochromatic incident beam is focused on a flat workpiece with a parallel grain pattern, such as surface-ground steel or sanded aluminum, the scattered light forms a diffraction pattern consisting of an irregular bar pattern parallel to the surface grain, which can be observed on a viewing screen facing the illuminated area. As the focusing is improved, the bar pattern becomes more widely spaced due to the decreasing spot diameter. If the workpiece is moved tangentially across the grain, the diffraction pattern moves across the viewing screen in the same direction as the workpiece when the incident lens is out of focus in one direction, and the diffraction pattern moves across the viewing screen in the opposite direction when the incident lens is out of focus in the other direction. At proper focus, if the workpiece is moved tangentially across the grain, the diffraction pattern gives the impression of similar superimposed patterns moving in opposite directions or a stationary pattern which changes shape without overall movement. This effect is clearly apparent even with multiple longitudinal mode helium-neon lasers, but it may not be apparent with the wider spectral bandwidths and multiple spatial modes of some solid-state lasers. Of course, a suitable infrared viewer must be used for infrared lasers.

The photodetector lens must be focused after the incident beam is properly focused. The photodetector optics must be adjusted to maintain the image of scattered light from the workpiece at the proper position on the photodetector, usually the center null point, and the photodetector lens must be focused. Focusing of the photodetector lens must be observed directly on the photodetector or image splitter, which usually requires some form of magnifier or microscope. Observation of the actual photodetector surface may be difficult, and the technique illustrated in Figure 4.26, which was not experimentally used, is suggested.

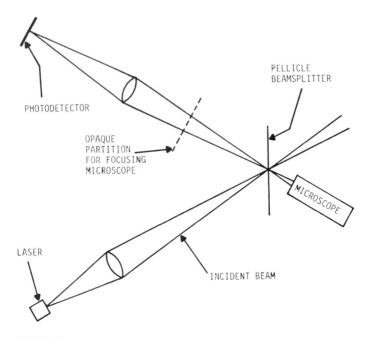

FIGURE 4.26   Possible arrangement for focusing the photodetector lens.

    In Figure 4.26, after the incident beam is properly focused
and the photodetector is focused as well as possible, a pellicle
beamsplitter is substituted for the workpiece surface. After
positioning the beamsplitter, the photodetector light path is
blocked and the microscope is focused on the beamsplitter by
observing scatter from the intense incident light. After the
microscope is focused, the photodetector light path is opened,
and focusing of the photodetector lens can be observed. If
the photodetector lens is properly focused, the focused incident
light reflected from the beamsplitter is focused on the photodetector, and reflected light from the photodetector is focused
back on the beamsplitter by reciprocity, which can be observed
through the beamsplitter with the focused microscope. If the
photodetector lens is not focused, the focused incident light
reflected from the beamsplitter is not focused back on the beamsplitter at the focal point of the microscope, and the improper
focus can be observed. The optical properties of the beamsplitter must be selected for usable results. A certain amount
of intentional nonspecular scatter may be required.

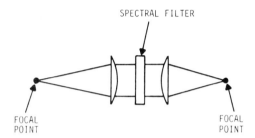

FIGURE 4.27   Using a collimated beam with a spectral filter in a focusing system to eliminate angular dependence.

The technique illustrated in Figure 4.26 is intended mainly for the linear position sensor, which responds to the true centroid of the incident light regardless of the image size. With a split-image photodetector or a photodiode array, the extent of the image on the photodetector can be observed by measuring the transition of the photodetector difference current or the output current of the central element of the array, respectively, as the distance to the workpiece is changed. The lens is focused to obtain the smallest image size and the sharpest transition. To avoid errors due to nonuniform reflectivity of surface grain, a mirror surface should be substituted for the workpiece after the incident beam is properly focused.

The transmission wavelength of a narrowband dielectric spectral filter depends on the angle of the light, and the characteristics are specified for perpendicular transmission. When the spectral filter is used in a focused beam, the angle of the light changes away from the center, and much of the light away from the center may be attenuated due to the narrow bandwidth. A very narrow filter bandwidth is not really necessary for attenuation of wideband ambient light, but most spectral filters inherently tend to have very narrow bandwidths. Such a filter can be used with a double lens, with the filter in the collimated beam between the elements, as illustrated in Figure 4.27. Of course, the very narrow usable filter bandwidth reduces the effects of ambient light.

A practical machining environment, where coolant is used, is a hostile environment for precision optical equipment. Occasional droplets of coolant and/or wet chips on the lenses are inevitable, and liquids on the lenses have adverse effects on

optical performance. Dry chips do not contaminate the lens
and mechanical damage to the lens surface is unlikely, but it
is desirable to divert all foreign materials, including dust,
from the lenses.

Foreign materials can be diverted from the lenses with a
forceful stream of clean, dry air emanating from the lens open-
ings. This requires a large volume of air at low pressure.
The use of standard compressed air lines is inefficient for this
purpose, and a suitable compressor, such as a vane pump,
should be used. A refrigeration-type air drier should be used
to dry the air to prevent contamination of the lenses.

Bulk movement of the air obviously does not affect the light
beams. Temperature gradients in still air affect the optical path
and the shape of the light beams due to random localized changes
in the index of refraction, and brisk movement of the air in the
optical path reduces temperature gradients.

Smoke from machining obviously affects the optical measurement. The forceful stream of clean, dry air from the lens open-
ings discussed above can serve to expel smoke from the optical
path and reduce temperature gradients. An auxiliary blower
for clear air may also be used to maintain a clear optical path.

The balance of the considerations that have to do with implementing the gauge are electronic in nature. These considerations are related to the design of the signal processing, shield-
ing, and selection of electronic components, keeping in mind
the low input currents from the photodetector and the signal
bandwidths involved. Suffice it to say that good electronic
practice should be followed to achieve the measurement accuracies stated.

CONCLUSION

Characteristics of a complete optical gauging system and machine
tool controls have been examined and analyzed. Applications to
other tasks are fairly obvious.

The noncontact gauging technique described is suitable for
rapid, accurate measurement of moving machined surfaces.
There are radar (time-of-flight) techniques for measuring large
distances with good relative accuracy and there are also interferometric techniques for very accurate distance measurement,

but they require good mirror reflectors, so they are not suitable for direct measurement of arbitrary surfaces.   The optical
gauging system described here bridges this gap and provides
an accurate method of measuring short distances to a nonspecular surface.

When this system is applied to numerical machine tool control, the machining errors can be enclosed within the feedback
loop, and the controlled dimension of the workpiece is determined solely by the accuracy of the optical gauging system.   A
unique advantage of this concept is that the cutting tool need
not be precisely located in relation to the workpiece.   Because
of this, a machining system does not require the expense and
complexity of preset tooling.   As mentioned previously, this
optical gauging principle can be scaled and adapted to a variety of machining and inspection applications, automated assembly, and robotic devices.

5

In-Process Inspection with Air Gauges

ROBERT A. THOMPSON / General Electric Company,
Schenectady, New York

INTRODUCTION

This chapter describes the development of an in-process inspection system for turning centers (lathes). Many such systems are under investigation, the technology having evolved along two paths, namely static and dynamic in-process inspection. In the static case, measurements are made between cuts, whereas dynamic systems measure and control on the fly.

The system described here represents General Electric's approach to dynamic in-process inspection. It uses an accurate displacement sensor adjacent to the cutting point to supply information for automatic tool condition compensation and automatic inspect/rework machine control. The work was performed under contract to the Air Force (Ref. 1).

This chapter puts special emphasis on the system's application to automated tool wear compensation. It outlines the development of the tool wear compensation adaptive control, especially its initial laboratory evaluation. It is broken into the following sections: overview of in-process inspection needs and systems, sensor development for General Electric's dynamic in-process inspection system, General Electric 1050T numerical control software modification for automatic tool wear compensation including

the interface between the sensor and control, and laboratory
demonstration of working automatic tool wear compensation control on the vertical turret lathe at GE's Aircraft Engine Business Group's machining development laboratory.

OVERVIEW

High labor and inventory costs combined with foreign competition are forcing U.S. industry to improve productivity. This
is the source of considerable incentives to automate manufacturing processes. On the other side of the issue, sophisticated
machine tool controls can now perform the complex control algorithms needed for process optimization. The need combined
with the potential capability of computer control is the key to
the current push for machine tool automation.

The main focus of the current automation effort has been
on in-process inspection. In-process inspection has as its major goal bringing parts to within acceptable tolerance limits without removing them from the machine. The savings which can
be achieved if this is done are obviously sizable because inspection on a separate machine and refixturing for part rework are
very costly. Of course, if on-line inspection could supply input
to automatic rework software, parts could be brought to within
tolerance without the intervention of an operator and truly un-
manned machining would result. Here the savings would be
outstanding.

Several systems have been conceived which can in theory
automatically perform the inspect/rework function. It is significant that in each of them the sensor is the most critical element. Part shape, once accurately measured, can be analyzed
and converted into necessary rework commands in the numerical
control without unreasonable technical difficulty.

Two systems approaches have emerged for in-process inspection, namely static and dynamic. In the static mode measurements are made in the absence of cutting. The numerical control is directed to measure the part surface relative to a known
reference in specified discrete locations.

Control logic computes errors and subsequent tool offsets
and the part is remachined. In this approach there are
several sensor schemes, including touch trigger probes, laser

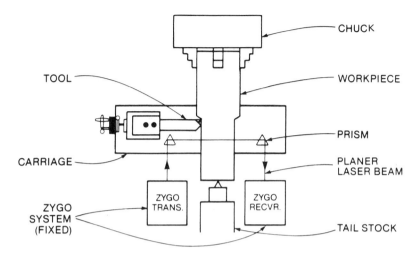

FIGURE 5.1   TRW's approach to in-process diameter measurements.

interferometers, and vibration touch probes. The static system
also lends itself to measuring tool wear by touching a known
reference and comparing with a previous touch.

In the dynamic approach measurements and correction are
made during the chip-forming process. Two dynamic approaches
have received attention under the Air Force contract. One by
TRW (Ref. 2) uses laser optical calipers as illustrated in Figure
5.1 to measure part diameter on the fly. The other is the subject of this chapter. It uses a displacement sensor in the tool
to compensate for tool condition during the cut and to perform
the same inspect/rework functions as the static sensors, but on
a continuous rather than discrete basis. To clarify the different approaches to in-process inspection, Table 5.1 summarizes
four existing systems. The table is by no means complete. It
covers only the systems dealt with under the aforementioned
Air Force contract, but it gives the reader an idea of how this
technology is evolving.

TABLE 5.1 Primary Characteristics of the Pratt & Whitney and Man Tech In-Process Inspection Systems

	Pratt & Whitney interferometer system	AEBG tool touch auditron	CRD air gauge	TRW Zygo optical calipers
System type	Static (spindle stopped when measuring)	Static (spindle running when measuring)	Dynamic (tool condition monitoring while machining)	Dynamic (finished part diameter measurement while machining)
Sensor type	Stylus, contact	Tool, contact	Air (or other) displacement gauge, noncontacting	Laser optical caliper, noncontacting
Measurement method	HeNe laser interferometer measures radial distance to work surface relative to gauge ring on spindle	Tool Touch Auditron stops feed when tool touches part. Resolvers on scales indicate coordinates relative to a fixed reference	Measures distance from a fixed point on the toolholder to freshly cut workpiece surface. Mastered against reference block	Measures shadow cast by workpiece to indicate part diameter as a function of the z-axis location
Measurement range	Exceeds swing of lathe in x-direction, covers full z range	Full range of machine slides	Full range of machine slides with 10 to 15 mil dynamic noncontacting range	In present development to 2 in. diameter
Type of measurement; adaptability to part geometry	OD or ID radius.[a] Not developed for contours	Unlimited by part geometry—can handle OD and ID contours	Can measure in any fixed OD or ID direction and can handle contours with tangents up to ±15°	All OD contours. Doesn't do IDs. Some accuracy loss with very steep tapers
Implementation requirements	Substantial—requires one turret tool location, laser and interferometer mountings, and a spindle-mounted gauge ring	Minimal—accelerometer on lathe turret is the only machine modification required	Modest—requires small air nozzle on tool with air supply	Substantial—requires mirror on z-axis carriage and accurately located laser sending and receiving units on machine frame
Uses	Part inspection—static tool condition	Part inspection—static tool condition	Part inspection—dynamic tool condition	Dynamic finished part inspection
Overall accuracy	Best	Very good	Good	Very good
Relative cost	Substantial	Minimal	Minimum	Modest
Main advantage	High accuracy—independent of machine position system	Adaptability to part geometry	Simplicity—broad range of applications	Ability to measure finished diameter on the fly

[a]OD, outer diameter; ID, inner diameter.

THE DYNAMIC APPROACH AT GENERAL ELECTRIC

Basic Approach

The dynamic in-process inspection system at General Electric Corporate Research and Development involves a displacement sensor mounted in the toolholder. Figure 5.2 illustrates the uses of a tool-mounted displacement probe. During a chip-forming operation, the device is used as a tool condition monitor. According to the figure, the device measures ε, the relative distance from a point on the toolholder to the freshly machined workpiece surface.

As the tool wears, it is clear that ε decreases an amount equal to the wear. This can be fed to the numerical control, which offsets the tool position to compensate for the wear. If the tool breaks, ε changes abruptly. Software can interpret this change as breakage and shut the machine down. It can also respond when ε reaches a preset minimum value indicating that the tool is worn out.

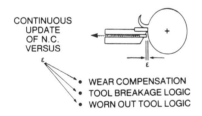

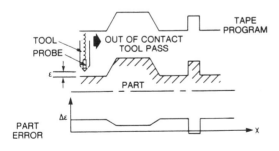

FIGURE 5.2 In-process inspection uses. (Top) Tool condition monitor. (Bottom) Finished part inspection.

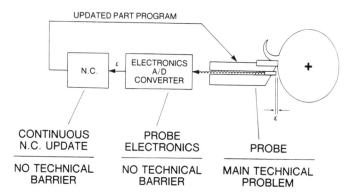

FIGURE 5.3   In-process inspection—technical barriers.

The second use of the tool-mounted displacement probe is also illustrated in Figure 5.2; it is the part inspection mode. It is important because when a part program directs the path of the tool during a cutting operation, the control has no way of knowing about and therefore cannot compensate for deflections in the machining system due to cutting forces, wear, etc. For example, as the tool wears and the cutting forces increase, the work is forced away from the tool and the actual depth of cut is less than the programmed depth by the combined effect of the wear itself and the wear-induced force increase.   The same effect arises in thin-walled or long slender workpieces which are subject to deflection during a cutting pass, and since the deflections are not compensated, they result in part errors.

The best way to evaluate this error is illustrated in Figure 5.2.   The tool is pulled a few mils out of contact with the part to eliminate the cutting forces.   The tape program is then run while monitoring ε.   If ε remains constant during this pass, it indicates that the part is identical to the program and no corrective action is needed.   If ε varies, the variation indicates the difference between the part and the program.   This difference can be fed into the numerical control on a block-by-block basis, and the modified part program can be rerun.   Through iterative inspect/rework cycles, the part can be brought into tolerance without the intervention of an operator.   From these comments it is clear that a reliable tool-mounted probe has several valuable in-process inspection and control applications.

The technical barriers to the implementation of controls based on such a probe therefore have to be addressed.

Technical Barriers to In-Process Inspection Using a Tool-Mounted Displacement Sensor

Figure 5.3 shows the basic control loop for the tool-mounted displacement probe in-process inspection system. The displacement signal from the probe is fed into electronics to convert it into an analog voltage. The analog voltage is converted into a digital indication of probe gap ε, after which it is sent to the numerical control, where a control strategy is executed and modified part program commands are fed back to the tool position drives.

A closer examination of the three basic functions described in Figure 5.3 indicates clearly the technical barriers which must be overcome to implement in-process inspection and tool condition compensation based on the probe signal.

Starting at the left of the figure, with the advent of microprocessor-based numerical controls the capability to do signal processing and control logic within the control is available. Therefore, although programming a control like the General Electric 1050T control to perform functions such as tool condition compensation is formidable, it is not a technical barrier. Certain software functions can be bypassed on a continuous real-time basis so that in-process position command modifications can be made. The probe electronics is also not a technical barrier, nor is the analog-to-digital (A/D) conversion. The probe electronics is a support function for the probe system. It simply converts probe information into an analog voltage.

The probe itself is a different matter; it must have a reasonable range and good sensitivity to achieve the accuracy needed for in-process inspection. Part accuracies of 0.0254 mm (0.001 in.) or better demand that the probe be able to resolve 0.00254 mm (0.0001 in.). It must be able to operate in the hostile environment of hot chips and flood coolant without error. The probe head must be simple and easily changeable in case of damage.

Furthermore, the probe head must be small so that it can be located near the cutting tip. There are two reasons for

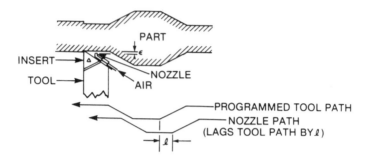

FIGURE 5.4   Tool modification with probe lagging cut by distance 1.

this requirement.   Figures 5.4 and 5.5 illustrate the first of these reasons.   Figure 5.4 shows a possible arrangement with the probe lagging the tool by the distance ℓ.   Here the center of the face of the probe is adjacent to the workpiece center so that small vertical deflections of the tool and probe are not sensed.   If ℓ is too great, however, the probe will run into the part when it is running down a contour or it will run out of range if it is running up the contour.   The lag ℓ can also introduce time delay problems in a control loop.   To avoid the problem associated with the lag ℓ, one might consider putting the probe under the cutting point as illustrated in Figure 5.5.   Another problem arises in this case, however.   In this position there is a variable offset ε_0 due to the curvature of the part.   Using the nomenclature of Figure 5.5, this offset goes as

$$\varepsilon_o = \frac{\delta^2}{D} \tag{5.1}$$

where D is the part diameter and δ is the distance from the top of the cutting insert to the probe center.   Due to tool deflections under the action of the cutting force, δ has a dynamic component along with its unloaded static component.   If δ = 6.35 mm (0.25 in.) and D = 2.54 cm (1.0 in.), the static offset is 1.60 mm (0.063 in.).   This establishes an order of magnitude for the range of the probe.

However, as smaller probes are used, δ, and with it the offset δ_0, decreases.   In aircraft propulsion components, D is large and the effect of the offset is reduced considerably.

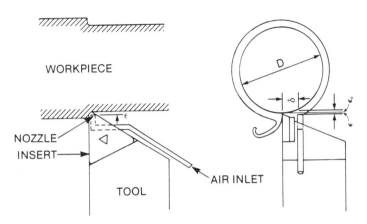

FIGURE 5.5 Tool modification with probe under the cutting point showing offset ε_o.

Thus it can be concluded that small probes allow the probe to be located near the cutting interface so as to reduce the problems suggested by Figures 5.4 and 5.5.

The second reason for specifying a small probe has to do with the integrating effect of the probe's face. That is, the cross feed in most machining operations is generally from about 0.152 mm (0.060 in.) per revolution, hence the workpiece must turn 10 revolutions before the probe fully covers the area machined. It would be advantageous to reduce this time-delayed effect with a smaller nozzle, say of 0.76 mm (0.030 in.).

Another very important consideration which arises in the inspection of contoured surfaces is the range-to-diameter ratio of the device. If the probe has a spherical head it is insensitive to the direction of the target, but most probes have flat heads. In these cases there is a directional offset ε_d which is nonzero when the target is not normal to the probe. Figure 5.6 shows ε_d as a function of the angle to the target. Since the probe measures the average distance to the target, it is clear that ε_d is given by the equation

$$\varepsilon_d = \frac{d}{2} \tan \theta \qquad\qquad (5.2)$$

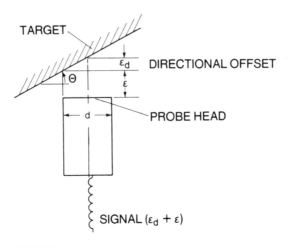

FIGURE 5.6 Directional offset ε_d versus target angle θ and probe diameter d.

where d is the diameter of the probe face and θ is the inclination of the gauged surface. Since the numerical control has information at all times on the angle θ, simple software can compensate for ε_d and give the true distance from the probe center to the target. The same steps can also be performed to compensate ε_o of Eq. 5.1, because the numerical control also carries continuous information on part diameter, D.

Equation 5.2 can be rearranged and written in terms of the range-to-diameter ratio of the probe, or

$$\text{Range-to-diameter ratio} = \frac{\varepsilon_d}{d} = 0.5 \tan\theta \qquad (5.3)$$

Thus, if the gauge is to operate so as to cover a full 90-degree contouring range, the maximum angle θ must be ±45 degrees and according to Eq. 5.3 the range-to-diameter ratio of the probe must be at least $\varepsilon_o/d = 0.5$. This, however, is a limiting minimum range, and a significantly larger range is needed to give a sufficient useful dynamic measuring range.

With these arguments in mind, it appears that the ideal probe would have the specifications shown in Table 5.2. In the next section it will be shown that these requirements cannot all be met simultaneously and that certain compromises have

TABLE 5.2 Ideal Specifications for a Dynamic In-Process
Inspection Sensor

Resolution	0.00254 mm (0.0001 in.)
Range	0.76 mm (0.030 in.)
Face diameter	0.76 mm (0.030 in.)
Range-to-diameter ratio	1
Environment	Must operate in hostile environment

to be made. This fact establishes a technical barrier for in-
process inspection which must be overcome.

REVIEW OF DISPLACEMENT SENSORS

An extensive review of noncontacting displacement sensors was
conducted to select the most appropriate sensor for the in-process inspection described in this chapter. The sensors investigated included the following:

1. Magnetic—reluctance and eddy current
2. Electrical—capacitance
3. Optical—fiber optic lever, laser imaging, and interferometric
4. Acoustic—phase shift and sonar
5. Pneumatic—air gauge

Of these choices, the air gauge was selected for dynamic in-
process inspection as being the most appropriate gauge. The
discussion which follows will clearly show why this is true.

Magnetic—Reluctance and Eddy Current

Reluctance

This sensor essentially measures the reluctance of the magnetic
loop formed by the probe and the measured surface. This reluctance increases as the probe gap increases. Because of its
operating principle, this sensor works only on ferromagnetic
materials. It is not practical for in-process inspection because
the ferromagnetic properties of most steels are affected by machining and because many superalloy materials are nonferromag-
netic.

Eddy Current

This probe transmits a high-frequency magnetic field and measures the effect on probe inductance of eddy currents generated
in the gauged surface. It works only on conducting materials
and is sensitive to the bulk resistivity of these materials. It
must, therefore, be calibrated for each alloy against which it
works, a decided disadvantage in machining applications. Probes
of this type have been developed to a high degree by Kaman
Scientific Company. They have good linearity and frequency
response (to 50 kHz), and the range for a probe 2.032 mm
(0.80 in.) in diameter is 0.254 mm (0.010-in. range). They
were considered for in-process inspection but were not used
because of their sensitivity to gauged surface material properties and because they lack the toughness of the air gauge.
Chips could also present a problem for this type of gauge.

This gauge may have a distinct advantage over the air
gauge in that, if equipped with a spherical head, it may work
for contoured surfaces. When the in-process inspection program has progressed to the point of treating contours, this
sensor will have to be considered.

Electrical—Capacitance

Since the capacitance between a conducting plate and a surface
is inversely proportional to their spacing, their separation can
be very accurately measured in terms of capacitance. This
gauge was initially thought to be a very good candidate for in-
process inspection because it has the desired range and sensitivity and because the instrument is insensitive to the properties of the material it is gauging. During initial tests it had
some problems related to liquids in the probe gap and electrical
shorting. These were overcome by putting a high-pressure air
jet down the center of the probe. However, a more fundamental problem with the capacitance gauge undermines its use for
in-process inspection. The problem is the size of the gauge
needed to achieve a given range. For example, a gauge with
a 0.254 mm (0.010 in.) range has a gauging head diameter of
6.35 mm (0.25 in.). This computes to an unacceptable range-
to-diameter ratio of 0.040, far too small in light of the specifications of the last section.

Optical—Fiber Optic Lever

Most optical gauges are inappropriate because of the dirt and harsh environment associated with machining and because they are sensitive to the reflectivity of the gauged surface. However, the fiber optic lever may have some merit for in-process inspection, so a brief description of its operation follows.

The fiber optic lever can best be explained in terms of two adjacent optical fibers, one a light transmitter and the other a receiver of reflected light, the amplitude of which is sensed by a photodiode. It is characteristic of optical fibers that they emit a cone of light rather than a parallel beam. When the probe is very close to the gauged surface, the cones of the sending and receiving fibers do not overlap and, consequently, no light is sensed by the receiving fiber. As the probe is drawn away from the surface, the cones overlap to a greater degree and the amount of received light increases, giving a measure of the probe gap.

In real applications, bundles of optical fibers optimize the response of this system. As would be expected, this system has a very high frequency response. The range per unit probe diameter, however, is no better than that of the air gauge.

Acoustic—Phase Shift and Sonar

Phase Shift

This probe measures the phase angle between a sent and a received sound wave, whose wave length is about 0.254 mm (0.01 in.) for a 1-MHz signal. This system is impractical for in-process inspection because of its complicated transmitter and receiver and because of the dispersive nature of the cutting environment with chips and coolant.

Sonar

This well-known system uses the time of travel for a sent and a received sound burst, coupled with a knowledge of the velocity of sound, to measure distance. It has the same disadvantages for a machining application as the phase shift type.

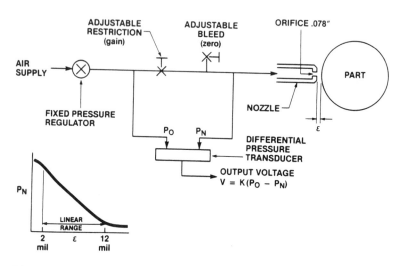

FIGURE 5.7 Edmunds air gauge.

Pneumatic—Air Gauge

There are two types of air gauges: the passive type, which is
characterized by a relatively short range, and the active type,
which has a much greater range.

The operating principle of the Edmunds Company's passive
air gauge is illustrated in Figure 5.7 (Ref. 3). The Edmunds
TRENDSETTER air gauge is used because it has an electrical
output, making it suitable to interface to an NC controller.
This section focuses on the characteristics of the passive air
gauge which make it appropriate for in-process inspection. It
also indicates why the active air gauge is not applicable.

The basic characteristics of the passive air gauge are ideal
for operating in the machining environment. Unlike other sensors such as the capacitance gauge, which has an electrically
charged head, or the eddy current probe, which has a current-
carrying coil, the air gauge has no complex components in the
chip-forming area. The probe is a simple, small air nozzle connected to the remotely located sensing unit by an air hose.

The displacement signal is resolved in the sensing unit from the back pressure in the hose. The nozzle can be operated at relatively high pressures, e.g., 0.586 MPa (85 psi), so that coolant and dirt are flushed from the gauging area. It has the further advantage that the nozzle is very inexpensive and easily replaced.

The equation for the maximum practical range of an air gauge is

$$\varepsilon_{max} = \frac{d}{4} \qquad (5.4)$$

where d is the diameter of the air gauge orifice. Since the smallest possible nozzle is desired, if the nozzle diameter is chosen equal to that of the smallest available magnetic eddy current gauge, 2.03 mm (0.8 in.), the air gauge has a maximum range of 0.508 mm (0.020 in.). This is double the available range for the eddy current gauge and it is achieved in a sensor which does not require any electrical connections. It results in a range-to-diameter ratio of 0.25, about the largest value obtainable in commercially available passive displacement probes. This fact and the fact that it is inexpensive and can tolerate the environment of the cutting process make the air gauge by far the most suitable for in-process inspection.

Even though the passive air gauge is better than any other commercially available gauge type, it still does not meet the ideal specifications given in Table 5.2. It has the desired 0.00254 mm (0.0001 in.) resolution and can operate in the machining environment, but it does not meet the desired range-to-diameter ratio of 1. It doesn't even meet the minimum range for contouring of $\varepsilon_d/d = 0.5$.

To overcome this problem one might consider active air gauges, which are capable of much greater range. In active air gauges the back pressure in the nozzle line is fed back to a piston connected to the nozzle so that when the probe gap increases this pressure drops, lowering the retracting pressure on the piston. This allows the nozzle to move closer to the gauged surface until pressure equilibrium is restored. In this way the device maintains a constant air gap over a large range of nozzle positions. A linear variable differential transformer (LVDT) senses the nozzle position and gives an electrical output with a resolution of 0.00254 mm (0.0001 in.).

The B.C. Ames Company has developed the active air gauge. For a typical gauge, the nozzle diameter is d (d_{approx}) 1.27 mm (0.050 in.) and the device is specified to operate at 0.101 mm (0.004 in.) from the target. This amounts to an effective range-to diameter ratio of $\varepsilon_d/d = 0.08$, although it has a large actual range (approximately 2.54 cm, or 1.0 in.).

For this gauge the maximum contouring angle can be calculated from Eq. 5.4 as $0.08 = 0.5 \tan \theta$ or $\theta = \pm 9$ degrees. Thus if the work surface was angled at more than 9 degrees relative to the probe, the nozzle would rub against the work. Consequently, the large range available with the active air gauge does nothing to solve the contouring problem, and its added complexity makes it undesirable as a sensor for in-process inspection.

This summary of displacement probes clearly points to the passive air gauge as the best sensor for General Electric's dynamic in-process inspection system, even though it does not fully meet the desired specifications. The range-to-diameter ratio needed for measuring ±45-degree contours cannot be attained by the air gauge or any other available gauge, so this problem will have to be solved by some other means, such as multiple nozzles.

INITIAL LABORATORY EVALUATION OF THE AIR GAUGE

This section summarizes laboratory evaluation of the air gauge, showing how it responds to tool wear, breakage, etc. under different laboratory conditions. The air gauge configuration for the whole test had the nozzle at the level of the triangular insert tip as shown in Figure 5.8. Reference 4 gives details of the behavior of the air gauge in sensing tool wear and breakage.

Wear Measurement and Inspection on a Manual Lathe

Figure 5.9 shows the wear behavior of a 370 grade tungsten carbide insert cutting 347 stainless steel. A 36.83 cm (14.5 in.) Monarch lathe with 137.16 cm (54 in.) centers was used to machine the straight shaft of about 10 cm (4 in.) diameter

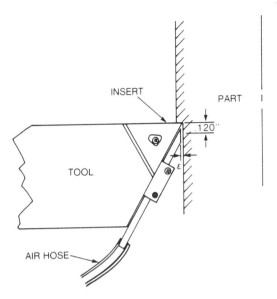

FIGURE 5.8   Air gauge instrumentation.

and 61 cm (24 in.) cut along the shaft. One end of the shaft
was rigidly supported in a three-jaw chuck and the other end
was supported by a live center. The live center end was considerably more flexible than the chuck end.

Figure 5.9 consists of two chart recordings. The lower
chart shows the results of an inspection pass run with the tool
and work out of contact after recording the upper chart. The
upper recording was made during a cutting pass. The upper
chart shows how the tool wore during the cut. The conditions
for the cut, such as depth of cut, workpiece material, and cutting speed, are shown on the figure to the left of the second
chart recording. The cut started at the left-hand end of the
chart, corresponding to the live center end of the workpiece,
and proceeded toward the right of the chart. The cutting
speed for the test was quite high at about 251 m/min, or 825
surface feet per minute (SFPM) and consequently the tool tip
wear rate was relatively high.

It is clear from the chart that at the start of the cut, the
initial air probe gap was about 0.18 mm (0.0072 in.) and after
some initial rapid break-in, wear proceeded at a relatively linear

156

Thompson

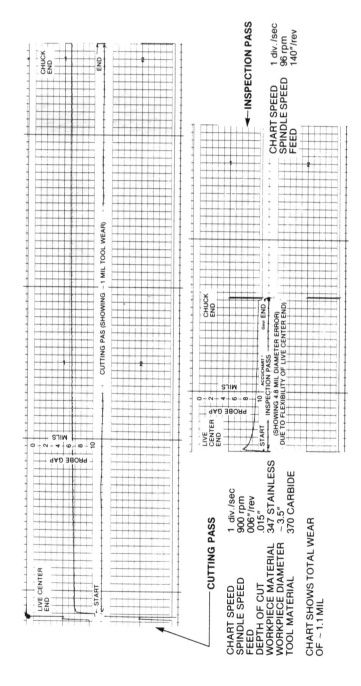

FIGURE 5.9 Demonstration of air gauge—tool condition.

rate for the remainder of the cut. At the completion of the cut, about 6 minutes later, the gap had decreased to about 0.15 mm (0.0061 in.), corresponding to an insert nose wear of 0.03 mm (0.0011 in.). The insert was measured by micrometer before and after the cut, and the measurement agreed very well with the wear indicated by the Edmunds air gauge.

It can be seen from the plot that the air gauge output signal-to-noise ratio was very good. It should also be noted that, in the absence of all other effects, wear would have caused a variation in shaft diameter of 0.06 mm (0.0022 in.), the live center end attaining the smallest diameter.

Next, consider the second chart recording of Figure 5.9. It shows the results of the tool/workpiece out-of-contact inspection pass for the feeds and speeds shown to the right of the chart. As with the first chart, the measurement started at the live center end and proceeded to the right to the chuck end. The chart shows an initial probe gap of 0.10 mm (0.0074 in.), which increased to 0.25 mm (0.0098 in.) at the completion of the pass. Consequently, the live center end diameter was 0.12 mm (0.0048 in.) larger than the chuck end diameter. This was verified by a micrometer. Since the tool wear effect was 0.06 mm (0.0022 in.) in the opposite direction, in the absence of tool wear the diameter variation would have been 0.18 mm (0.0048 + 0.0022 = 0.0070 in.). The reason for this effect was the difference in machine flexibility between the live center and chuck ends. For this particular machine-tool-workpiece system, it overshadowed the wear effect. That is, the live center end deflected away from the tool more under the normal cutting forces than the chuck end. This effect resulted in a smaller depth of cut on the live center end, an effect which tended to leave a larger diameter on this end. It is interesting to note from the inspection curve that this effect was not linear along the length of the shaft. This can be attributed to the nonlinear deflection behavior of a beam under bending.

The dimensional error of the part illustrated by Figure 5.9 was dominated by the shaft's stiffness. However, in very high wear situations, such as high-speed machining of Inconel or titanium, the tool wear effect might dominate. Therefore, both effects are probably equally important and should be compensated by appropriate automatic NC logic. Many other tests like the one described here were run on a similar Lodge and Shipley manual lathe with the same results.

Tool Wear and Breakage in Superalloy Machining

Material for this subsection was taken from Ref. 4. Only a couple of figures from the report are presented here to show the Edmunds air gauge working on a different lathe under different conditions and to show the diagnostic capability of the device. The tests were run on the 63.5 cm (25 in.) Lodge and Shipley variable-speed lathe in the Materials Removal Laboratory at the Aircraft Engine Group, Cincinnati.

The object of the test was to evaluate the air gauge as a sensor for breakage. A displacement probe should work well as a breakage sensor because it measures the gap between a reference on the toolholder and the freshly machined surface. When breakage occurs, this gap closes impulsively.

Tool breakage detection by air gauging has several inherent advantages:

1. Since it measures the gap between the toolholder and the freshly cut work surface, the air gauge measures breakage directly rather than through secondary effects such as vibration and force.
2. Its output is a direct indication of breakage (impulsive gap closure); it does not require complex logic circuits for its function.
3. Since it is a pneumatic device, the air gauge has a time response of about 0.1 to 1 second. It therefore filters out high-frequency vibrational effects which ride on top of the breakage signal.

Several tests were performed on the Lodge and Shipley lathe. Although the major objective of the work was to detect tool breakage in ceramic inserts, two types of inserts and workpieces were evaluated: 883 tungsten carbide cutting titanium and triangular ceramic insets of 3.2 mm (1/8 in.) and 1.59 mm (1.16 in.) nose radius cutting Inconel 718. The tests of tungsten carbide cutting titanium showed the air gauge operating as a tool wear detector and unexpectedly uncovered a mechanical problem with the Lodge and Shipley lathe. The tests of ceramic inserts cutting Inconel 718 demonstrated the device as a tool breakage detector.

Before the cutting tests were started a question concerning the effect of flood coolant against the air nozzle was raised.

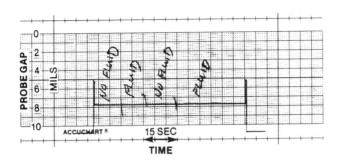

CHART SPEED	1 (SMALL) DIV/SEC
CUTTING SPEED	0 SFPM
DEPTH OF CUT	0
CROSS FEED	0
TOOL MATERIAL	1/8" NOSE RADIUS TRIANGULAR CERAMIC
WORKPIECE MATERIAL	6-4 TITANIUM
WORKPIECE DIMENSIONS	APPROX. 6" DIA. × 24" LONG

FIGURE 5.10   Effect of coolant flow on the Edmunds air gauge.

To evaluate the effect of cutting fluid, the spindle was stopped and the air gap was set arbitrarily at about 0.198 mm d (0.0078 in.). The coolant pump was turned on and the coolant stream was directed at the air nozzle. The chart recorder was started and the coolant flow was turned on and off while recording the air gauge output. Figure 5.10 shows the result. It is clear from the figure that coolant flow had no effect on the air gauge output. In fact, no change was observed on the Edmunds air gauge column readout, which is resolved into 0.00254 mm (0.0001 in.) divisions. Thus, coolant does not introduce an error into the air gauge readings.

With the coolant issue resolved, consider the first test run using a triangular 883 carbide tool with a titanium workpiece. These tests showed the dramatic effect of the normal cutting force increase associated with wear. They also showed the value of air gauging as a diagnostic technique for troubleshooting lathe errors. The first test in this series was run with the conditions listed in Table 5.3.

The cutting force was low and the tool wore very slowly. The cut in each case started at the live center end of the lathe and ended at the chuck end. Figure 5.11 shows the air gauge

TABLE 5.3   Cutting Conditions for the 883 Carbide/
Titanium Tests

Depth	1 mm (0.040 in.)
Surface speed	30.5 m/min (100 SFPM)
Feed	0.25 mm/rev (0.010 in./rev)

output at the start of the test and Figure 5.12 shows the air gap at its completion.   Things to be noted in these curves are, first, the 0.064 mm (0.0025 in.) once-per-revolution oscillation of the curve at the beginning of the cut (Figure 5.11); second, the smaller once-per-revolution oscillation of 0.038 mm (0.0015 in.) at the end of the cut; and third, the gap closure of 0.036 mm (0.0014 in.) at the end of the cut due to the decay of cutting force and the subsequent tool springback at the end of the cut.   According to Figures 5.11 and 5.12, the tool wore about 0.018 mm (0.0007 in.) during the cut of about 27.9 cm (11 in.) across the 6-4 titanium shaft.

As the next step, the tool was pulled out of contact with the work and another pass was made while monitoring the air gauge.   Figures 5.13 and 5.14 show the beginning and end of

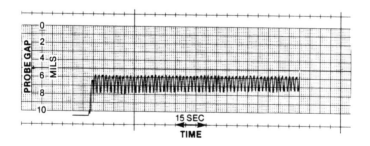

CHART SPEED	1 (SMALL) DIV/SEC
CUTTING SPEED	100 SFPM
DEPTH OF CUT	.040"
CROSS FEED	.010"/REV
TOOL MATERIAL	TRIANGULAR 883 CARBIDE
WORKPIECE MATERIAL	6-4 TITANIUM
WORKPIECE DIMENSIONS	APPROX. 5.6" DIA. × 24" LONG

FIGURE 5.11   Live center end (low-speed cut on titanium).

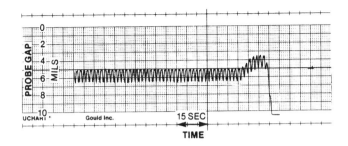

CHART SPEED	1 (SMALL) DIV/SEC
CUTTING SPEED	100 SFPM
DEPTH OF CUT	.040"
CROSS FEED	.010"/REV
TOOL MATERIAL	TRIANGULAR 883 CARBIDE
WORKPIECE MATERIAL	6-4 TITANIUM
WORKPIECE DIMENSIONS	APPROX. 5.6" DIA. × 24" LONG

FIGURE 5.12   Chuck end (low-speed cut on titanium).

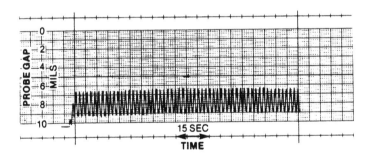

CHART SPEED	1 (SMALL) DIV/SEC
CUTTING SPEED	100 SFPM
DEPTH OF CUT	—
CROSS FEED	.010"/REV
TOOL MATERIAL	—
WORKPIECE MATERIAL	6-4 TITANIUM
WORKPIECE DIMENSIONS	APPROX. 5.6" DIA. × 24" LONG

FIGURE 5.13   Live center end (tool out of contact).

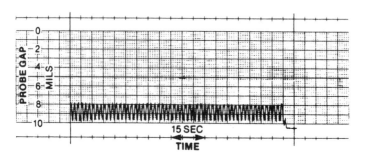

CHART SPEED	1 (SMALL) DIV/SEC
CUTTING SPEED	100 SFPM
DEPTH OF CUT	—
CROSS FEED	.010"/REV
TOOL MATERIAL	—
WORKPIECE MATERIAL	6-4 TITANIUM
WORKPIECE DIMENSIONS	APPROX. 5.6" DIA. × 24" LONG

FIGURE 5.14   Chuck end (tool out of contact).

this pass.   The initial minimum gap was 0.18 mm (0.007 in.) at
the beginning of the cut, and it grew to 0.20 mm (0.008 in.)
at the end of the pass.   This change in the DC level of the
probe output of the gap was very small and was in a sense opposite to the effect on diameter of tool wear.   The anomaly
amounted to only 0.043 mm (0.0017 in.) (0.018 mm or 0.0007 in.
wear = 0.025 mm or 0.001 in. due to inspection) and it can
probably be attributed to an alignment error of the machine
ways relative to the workpiece axis that amounted (if the cutting force was constant) to 0.043 mm (0.0017 in.) over the
length of the workpiece.

The variance of 0.043 mm (0.0017 in.) in the DC level of
the probe output was small.   More important was the large
once-per-revolution modulation of the probe gap during the cut
and the inspection pass.   It was 0.064 mm (0.0025 in.) at the
start of the cut and decreased to 0.038 mm (0.0015 in.) at the
end of the pass.   It was initially thought that this might represent a malfunction of the air gauge, so a dial indicator was
used to check the shaft.   The dial indicator agreed exactly in
all cases with the air gauge.   This led to a search for the
problem, which was shortly traced to the tail stock bearing.
The ball race precessed through approximately one complete
orbit during each two revolutions of the workpiece.   It was

TABLE 5.4 High-Wear Cutting Conditions for the 883
Carbide/Titanium Tests

Depth	1 mm (0.040 in.)
Surface speed	91 m/min (300 SFPM)
Feed	0.25 mm/rev (0.010 in./rev)

discovered that the modulation of the probe gap really had a
period of once per two revolutions instead of the once per revolution originally assumed. Thus the problem was attributed to
an undersized ball or deterioration of the ball cage of the tail
stock bearing. The bottom line was that the air gauge was a
very good diagnostic tool which turned up a problem that had
been and may otherwise have continued to be overlooked.

The last step in the 883 carbide/6-4 titanium test was to
evaluate the air gauge as a tool wear detector for this tool-
workpiece combination. To achieve significant amounts of wear,
the cutting speed was increased by a factor of 3 above that for
the test described above. Thus, a cut was made with the conditions given in Table 5.4.

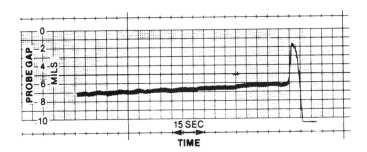

CHART SPEED	1 (SMALL) DIV/SEC
CUTTING SPEED	300 SFPM
DEPTH OF CUT	.040"
CROSS FEED	.010"/REV
TOOL MATERIAL	TRIANGULAR 883 CARBIDE
WORKPIECE MATERIAL	6-4 TITANIUM
WORKPIECE DIMENSIONS	APPROX. 5.5" DIA. × 24" LONG

FIGURE 5.15 Cutting pass (high-speed cut on titanium).

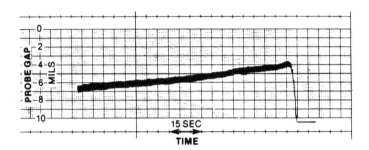

CHART SPEED   1 (SMALL) DIV/SEC
CUTTING SPEED   300 SFPM
DEPTH OF CUT   .040"
CROSS FEED   .010"/REV
TOOL MATERIAL   TRIANGULAR 883 CARBIDE
WORKPIECE MATERIAL   6-4 TITANIUM
WORKPIECE DIMENSIONS   APPROX. 5.5" DIA. × 24" LONG

FIGURE 5.16 Inspection pass (high-speed cut on titanium).

Since the workpiece was about 14 cm (5.5 in.) in diameter, the spindle speed for this test was about 210 rpm (or 3.5 rev/s), whereas in the previous test it was about 70 rpm (1.17 rev/s). Figure 5.15 shows the air gauge output for the last 100 s of the cut, while Figure 5.16 shows the inspection pass taken over the same portion of the workpiece with the tool and workpiece out of contact.

Before going into the aspects of tool wear for this test and its effect on the workpiece, attention should be given to the oscillation of the probe signal. These oscillations have been attributed to a bad bearing in the tail stock. At the end of the test run at 210 rpm, shown by Figures 5.15 and 5.16, their peak-to-valley amplitude amounted to about 0.013 mm (0.0005 in.), whereas at the end of the test run at 70 rpm, shown by Figures 5.12 and 5.14, the oscillations were 0.038 mm (0.0015 in.).

This effect was *not* due to a reduction of the shaft oscillation. Instead, it was due to the inherent filtering characteristic of the air gauge. Since the oscillations for the 70-rpm test were checked with a micrometer, it is clear that the speed of the air gauge response was somewhere between the limits set by these two tests, say 140 rpm. Next, it will be remembered that the bearing characteristic was such that the shaft oscillations were one-half the spindle speed. Thus the maximum frequency to which the instrument could respond was about 70 oscillations/min or approximately 1 Hz. This would give the device a nonattenuated response time of about 1 second. The response time is due mainly to the air column between the instrument and the air nozzle and could be shortened by shortening the air hose.

Figure 5.15 shows the tool wear for the last 100 seconds of the test. It is clear that during that period, the tool wear was such that the air gap decreased from about 0.18 mm (0.0072 in.) to 0.16 mm (0.0062 in.). The steep decline of the air gap to 0.04 mm (0.0016 in.) at the end of the cut was due to the loss of cutting force as the tool ran off the end of the work. The step shows the sizable springback of 0.12 mm (0.0046 in.) in the machine tool system due to cutting force.

Next, consider Figure 5.16. It shows the inspection pass with the tool and work out of contact for the part of the shaft shown by Figure 5.15. The gap during this inspection pass varied from 0.17 mm (0.0067 in.) to 0.1 mm (0.004 in.). This amounts to a diameter change of 0.14 mm (0.0054 in.). A micrometer check of the shaft showed the shaft dimensions to be 141.05 mm (5.553 in.) at the start of the section and 141.17 mm (5.558 in.) at its end, or a change of diameter of 0.12 mm (0.005 in.), which is in good agreement with the air gauge readings.

Now, since the tool wear for this section of the shaft was shown to be 0.025 mm (0.001 in.), only 0.05 mm (0.002 in.) of the diameter variation can be accounted for by wear. The remaining 0.086 mm (0.0034 in.) on diameter, or 0.043 mm (0.0017 in.) on radius, must be attributed to the cutting force buildup accompanying wear. This effect tended to force the work away from the tool and leave a larger diameter than would occur in the absence of the force change. The larger finished diameter in turn accounted for the smaller air gap during the inspection pass. This shows the sizable effect of cutting force on part geometry.

As stated, the ultimate objective of the test run on the 25-in. Lodge and Shipley lathe was to determine whether air gauging can reliably detect ceramic cutter breakage. After considerable experimentation with a ceramic insert and Inconel 718 workpiece combination, suitable conditions for the breakage tests were found to be those given in Table 5.5.

TABLE 5.5   Conditions for the Tool Breakage Detection Tests

Ceramic tool	1.59 mm (1/16 in.) nose radius
Depth	2 mm (0.080 in.)
Cutting speed	183 m/min (600 SFPM)
Feed	0.15 mm/min (0.006 in./rev)
Workpiece prep.	Chamfer

It was found that some chamfer was needed on the work-piece to prevent the tool from breaking on initial contact with the work.   After the need for a chamfer was established, sev-eral tests were run in which breakage occurred during the cut.

Figures 5.17, 5.18, 5.19, and 5.20 show the probe output for four of these tests.   In each of the four tests an abrupt drop of the probe gap to around 0.025 mm (0.001 in.) gave a positive indication of breakage.   The time to breakage for the tests varied from 7 seconds for Figure 5.18 to 25 seconds for Figure 5.20.   It is evident in each case that there were large force variations, due probably to chipping during each cut. Chipping caused the normal cutting force to increase or decrease,

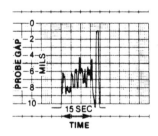

CHART SPEED	1 (SMALL) DIV/SEC
CUTTING SPEED	600 SFPM
DEPTH OF CUT	.080"
CROSS FEED	.006"/REV
TOOL MATERIAL	1/16" NOSE RADIUS TRIANGULAR CERAMIC
WORKPIECE MATERIAL	718 INCONEL
WORKPIECE DIMENSIONS	APPROX. 4.8" DIA. × 24" LONG

FIGURE 5.17   Test showing total breakage for cutting conditions indicated.

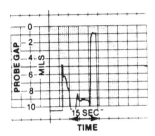

CHART SPEED	1 (SMALL) DIV/SEC
CUTTING SPEED	600 SFPM
DEPTH OF CUT	.080''
CROSS FEED	.006''/REV
TOOL MATERIAL	1/16" NOSE RADIUS TRIANGULAR CERAMIC
WORKPIECE MATERIAL	718 INCONEL
WORKPIECE DIMENSIONS	APPROX. 4.8'' DIA. × 24'' LONG

FIGURE 5.18   Test showing total breakage for cutting conditions indicated.

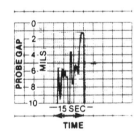

CHART SPEED	1 (SMALL) DIV/SEC
CUTTING SPEED	600 SFPM
DEPTH OF CUT	.080''
CROSS FEED	.006''/REV
TOOL MATERIAL	1/16" NOSE RADIUS TRIANGULAR CERAMIC
WORKPIECE MATERIAL	718 INCONEL
WORKPIECE DIMENSIONS	APPROX. 4.8'' DIA. × 24'' LONG

FIGURE 5.19   Test showing total breakage for cutting conditions indicated.

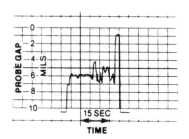

CHART SPEED	1 (SMALL) DIV/SEC
CUTTING SPEED	600 SFPM
DEPTH OF CUT	.080"
CROSS FEED	.006"/REV
TOOL MATERIAL	1/16" NOSE RADIUS TRIANGULAR CERAMIC
WORKPIECE MATERIAL	718 INCONEL
WORKPIECE DIMENSIONS	APPROX. 4.8" DIA. × 24" LONG

FIGURE 5.20 Test showing total breakage for cutting conditions indicated.

depending on whether it effectively sharpened or dulled the tool tip. The force change was reflected in the erratic variations of the probe gap as the work was forced away or allowed to approach the tool under the effect of these forces. Nose wear due to chipping, as evidenced by an appreciable decrease in the DC level of the probe gap, can be observed in Figures 5.17 and 5.19.

However, the initial indication of breakage varied from test to test because if breakage dulls the tool, the associated cutting force increase will initially open the gap, and vice versa if the break sharpens the insert. In each of the tests, although the gap initially opened, the amount that it opened varied widely from test to test. In each case, about 1 second elapsed between when the gap started to open after breakage and when it started to close. This is explained by the 1 second (i.e., 10 rev/s × 1 s × 0.15 mm or 0.006 in./rev = 1.5 mm or 0.06 in.) required for the nozzle to start looking at the surface generated by the broken tool. This effect can be reduced by decreasing the lag of the nozzle behind the cut—3 mm (0.120 in.) in Figure 5.8. A more complete explanation of the phenomena can be found in Ref. 4.

Tests on an NC Controlled Vertical
Turret Lathe

Several tests were run on a Gray vertical turret lathe. These
tests were run as part of the laboratory evaluation of automatic
tool wear compensation adaptive control. Since many of the
tests in this series were run uncompensated, they add to the
library of information reported in this section. However, it is
more appropriate to discuss them as part of the laboratory evaluation of the adaptive control.

AUTOMATIC TOOL WEAR COMPENSATION CONTROL

The automatic tool wear compensation system was developed for
the machine on which it was to be demonstrated, namely the
Gray vertical turret lathe (VTL) in the Machining Development
Laboratory of General Electric's Aircraft Engine Business Group
in Cincinnati. Specifications for this machine, which had a
General Electric 1050T numerical control, are given in the next
section.

Two subsystems were needed to close the automatic tool
wear compensation loop between the Edmunds air gauge and the
Gray VTL's x-axis controller board. They were (1) the interface between the air gauge and the 1050T control to provide
the control with continuous digital tool wear information and
(2) the 1050T software needed to convert tool wear information
into position commands for the machine's x-axis servo controller.
The next section describes the subsystems.

Air Gauge Interface to the 1050T Numerical
Control

Automatic compensation of part geometry for tool wear was controlled by software in the 1050T numerical control. Digital wear
information was received by the 1050T control from an A/D converter which was driven by the analog output of the Edmunds
air gauge. The A/D converter used was a model ADC-71C board,
which is an optional plug-in board for the General Electric 1050T
numerical control. This is a 10-bit board with several voltage
ranges, one of which was ±2.5 volts, and only half the board's
total available range of 512 bits were usable. However, this was

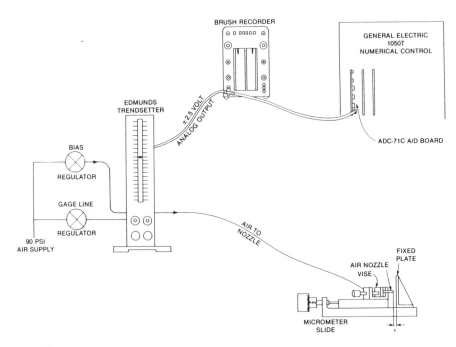

FIGURE 5.21   Air gauge interface to the 1050T numerical control and micrometer calibration.

more than adequate because only 100 of the remaining 256 bits were needed.   That is, although the 2 mm (0.080 in.) nozzle used had a maximum available range of 0.508 mm (0.020 in.), only 0.254 mm (0.010 in.) with a 0.00254 mm (0.0001 in.) sensitivity was important.   This was because the 1050T's position encoders had a 0.00254 mm (0.0001 in.) resolution and because the sensitivity of the micrometer table used to calibrate the system was also 0.00254 mm (0.0001 in.).

Thus, the air gauge was connected directly into the 1050T via the ADC-71C board as shown in Figure 5.21, and no intermediate signal conditioners or filters were needed.   The result was a very simple and trouble-free electronic package.

The system was calibrated with the aid of a micrometer table, also shown in Figure 5.21.   The table consisted of a vise into which the air nozzle could be clamped.   The vise was in turn mounted on a table which could be moved relative to a stationary plate by a very accurate 0.00254 mm (0.0001 in./division micrometer. Using this scheme, the air nozzle was accurately adjusted to a distance of 0.279 mm (0.011 in.) from the stationary plate. The micrometer screw was then used to move the nozzle in increments of 0.00254 mm (0.0001 in.) toward the plate. For each increment, the digital output of the A/D board to which the air gauge was connected was put in one column of a table in the 1050T's memory. At the same time, for each increment the corresponding air gap was determined from the micrometer screw and placed in the other column in appropriate digital form. This procedure was continued for 100 steps so that the resulting table output was the digital air gap at 0.00254 mm (0.0001 in.) steps between 0.0254 mm (0.001 in.) and 0.279 mm (0.011 in.).

At the same time that the digital table was being established, the output of the chart recorder and the Edmunds column readout were checked. Thus, at the completion of the calibrating procedure the system was completely compatible from the Trendsetter column gauge, to the recorder, to the air gap table in the 1050T control. This procedure for calibrating the system was both fast and effective, only about 20 minutes being required to set up the air gap table in the 1050T control and simultaneously calibrate the other instruments.

With the air gap data thus digitally available to the 1050T control, the wear control logic itself can be considered.

Continuous Closed-Loop Machining—Automatic
Tool Wear Compensation Software

Figure 5.22 shows the operator commands necessary to initiate continuous closed-loop machining (CCLM), applied specifically in this case to automatic tool wear compensation. From the figure it will be noticed that three steps are necessary. The first command initializes the system and orders it to position itself to take initial zero-wear data and subsequent wear offsets. The second triggers CCLM itself for a straight cut. It is broken into four commands. The first two of these (G01 and Z10.000) command a straight cut of 25.4 cm (10 in.) length; the third (M67) commands tool wear compensation, and the fourth (V100) orders the time interval between compensation steps. This interval equals the delay between the time when a point on the

I GO1   X01.000   M68

COMMANDS   WIPES OUT   TRIGGERS TO INITIATE
EXECUTION OF   PREVIOUS INFORMATION SYSTEM FOR AUTOMATIC
LINEAR MOTION   AND ORDERS X-ASIS   TOOL WEAR COMPENSATION
                  TO POSITION
                  ITSELF TO TAKE
                  INITIAL ZERO WEAR
                  DATA AND SUBSEQUENT
                  WEAR OFFSETS

II GO1   Z10.000   M67   V100

COMMANDS   COMMANDS   TRIGGERS WEAR   FREQUENCY OF
EXECUTION OF   LENGTH OF CUT   COMPENSATION   COMPENSATION
LINEAR MOTION   (10 INCHES)   EXECUTIVE SOFTWARE   (100 MILLISECONDS)
                                       AS WELL AS
                                       TOOL BREAKAGE
                                       FAILSAFE

III M69

WILL DISPLAY
TOTAL WEAR
COMPENSATION FOR
10 SECONDS AT
THE END OF THE RUN

FIGURE 5.22   Commands for operator execution of CCLM automatic tool wear compensation.

part is cut and when it is inspected by the trailing air nozzle.
More will be said later regarding this interval. Finally, the
third command displays the total tool wear for 10 seconds after
the completion of the run.

It will be noticed that a different M-code is used in each of
the commands. The most significant of these is M-67, which
triggers the necessary logic for tool wear compensation. This
brief software summary concentrates on M-67 and the software
it controls.

Figure 5.23 shows how a typical M-code command is processed through a General Electric 1050T numerical control. It
also shows the special features of CCLM. Block processing
sends processed command blocks to the axis complete interrupt,
which schedules M-code tasks, both premotion and postmotion,
and sends them to the schedule queue. These commands become
input to task executive, which executes the M-code task.

From task executive, the normal routing for a simple M-code
is through the output image area to the output boards, which
trigger lights, relays, etc. This normal path for simple M-codes

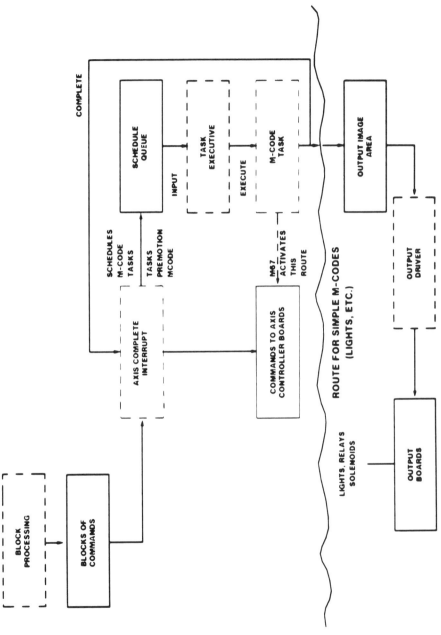

FIGURE 5.23 Continuous closed-loop machining (CCLM) software.

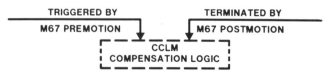

INITIAL LOGIC STEPS:

• DELAY 3 SECONDS (PROGRAMMABLE)
  AND CHECK FOR TERMINATION

• INITIAL INTERVAL

  - GET A/D CONVERTED VOLTAGE FROM AIR GAGE
  - COMPARE VOLTAGE TO PRE-CONSTRUCTED TABLE OF
    AIR GAP VERSUS GAGE VOLTAGE
  - ESTABLISH INITIAL AIR GAP = INITIAL VALUE
    FOR NEXT INTERVAL

FIGURE 5.24   CCLM tool wear compensation logic.

is shown below the wavy demarcation line of Figure 5.23.   The
complex M-code, M-67, for automatic tool wear compensation is
handled differently, entirely above the wavy line of Figure 5.23.

The CCLM software is housed in task executive.   This software calculates the offsets needed for automatic tool wear compensation.   But instead of sending the resulting information to
the axis controller boards via the normal routing along the "complete" path back to axis complete interrupt, the special feature
of the CCLM software is that the M-67 activates a direct route
from task executive to the axis controller boards.   Thus, the
response is prompt.   The details of the CCLM software triggered and terminated by the M-67 are shown in Figures 5.24
and 5.25.   Figure 5.24 illustrates the initial logic steps performed by this software, while Figure 5.25 shows subsequent
steps.

Following the initial logic steps, the first operation performed by the software is a time delay and check for termination of CCLM.   The time delay is basically the time required
for the tool to advance the distance between its trailing edge
and the back of the air nozzle.   The reason for the delay has
been explained before.   It is most easily understood for the
initial step by considering the start of a cut.   When the tool
starts to cut, the trailing air gauge is still off the work.   Then
after the tool proceeds the distance of the lag, the nozzle moves

SUBSEQUENT LOGIC STEPS:

- SUBSEQUENT 3 SECOND INTERVALS
  - GET A/D SIGNAL
  - CONVERT SIGNAL TO TABLE INDEX
  - PICK-UP NEW AIR GAP
  - IF NEW < OLD CALCULATE COMPENSATION = OLD - NEW
       IF COMPENSATION > MAXIMUM ALLOWABLE APPLY E-STOP
       PERFORM CCLM ADJUSTMENT BY SENDING ADJUSTMENT
       COMMAND TO AXIS CONTROLLER DIRECTLY
       CURRENT TOOL WEAR = TOOL WEAR ON LAST
       INTERVAL + COMPENSATION
       OLD VALUE = NEW VALUE
  - END IF

FIGURE 5.25   CCLM tool wear compensation logic.

onto the end of the work. After the nozzle completely covers
the end of the work, it measures the initial gap associated with
an unworn tool. The time delay ensures that the gap is not
measured until the nozzle is fully on the work. Figure 5.8 illustrates how the nozzle trails the tool tip. The software responds to this initial gap by getting the A/D-converted air
gauge signal from the ADC-71C board, comparing it with the
preprogrammed table of air gap versus this signal, and putting
the resulting value in memory. The memorized value thus obtained is the initial air gap, and it equals the comparison value
for the next interval. It represents the gap for an unworn tool.
The logic steps for the next and subsequent intervals are illustrated in Figure 5.25. In each case, they are performed following the programmed time delay, thus ensuring that the surface
machined following any tool position change is observed by the
air gauge before another change is made. According to the information in Figure 5.25, the first step in subsequent intervals
is to get the A/D-converted air gauge signal and, using the
table, convert it into a new air gap for that interval. If the
new gap is less, then, the gap equal to the difference is needed
to bring the tool tip back to its programmed position. This
compensation, in tenths of mils, is fed directly to the axis control boards through the path initiated by the M-67 (see Figure
5.23) and the new air gap is put in memory. This process is

continued until the total offset reaches a predetermined value which indicates that the tool is worn out or until some fail-safe in the software recognizes a problem and signals an E-stop (emergency stop). In the initial software, an E-stop is triggered if the calculated wear compensation value exceeds some preset maximum indicating tool breakage. Following the sequence of steps shown in Figure 5.25, an "END IF" recycles the logic. It is restarted by getting the A/D-converted signal following the programmed time delay; the logic steps are then retraced and the next compensation made.

LABORATORY EVALUATION OF AUTOMATIC TOOL WEAR COMPENSATION
The Test Setup

The conclusion of this chapter describes the actual closed-loop operation of the automatic tool wear compensation network. The tests were run at General Electric, AEBG, Evendale. The machine used, a Gray vertical turret lathe, is described in Table 5.6.

The test workpiece was a 6-4 titanium development spool (disk). Titanium works well for wear tests because it induces high wear in carbide inserts under elevated cutting speeds. The test part used for all the tests described in this section had the dimensions given in Table 5.7.

The Test Procedure

The basic test procedure and cutting conditions were not changed throughout the duration of the tests with the exception of the spindle speed, inspection interval, and insert type, which were the test variables. Thus, it is easy to compare the results. The test procedure involved writing a sample program for a straight cut over the 5.56 cm (2-3/16 in.) length of the development spool. This was fed into the machine by manual data input (MDI). The cutting conditions used throughout were as given in Table 5.8.

After establishing the machining conditions, the air gauge was connected to the 1050T's optional ADC-71C A/D board and the machine instrument interface was established. As a first step the air gauge calibration was checked using the setup

TABLE 5.6   Machine Tool Used to Evaluate Automatic Tool Wear Compensation[a]

Table drive	Motor specifications—75 hp at 650 rpm reversing Table rpm indicator and % load meter included Two speed ranges: high range, 12 to 300 rpm spindle speed; low range, 4 to 101.5 rpm spindle speed
Table specifications	1.219 m (48 in.) diameter—set of four 20.32 cm (8 in.) face plate jaws (for use with plain table)
Dimensions	Maximum height under rail head saddle—1.219 m (48 in.) Right hand, turret-type rail head—special 0.4572 m (18 in.) Power indexed tool block, four-sided Maximum height under turret block—1.372 m (54 in.) Feed, single range—infinitely variable 1.27 to 1524 mm/min (0.05 to 60 in./min) Rapid traverse rate—5.08 m/min (200 in./min)
Miscellaneous features	Chip conveyor Way covers on rail Refrigeration system Flood coolant system with distributor on right-hand head turret

[a]Gray VTL (vertical turret lathe) with General Electric 1050T numerical control—machine in-house at AEBG, Evendale.

TABLE 5.7   Workpiece for the Automatic Tool Wear Compensation Tests

6-4 titanium development spool
43.18 cm (17 in.) diameter
5.56 cm (2-3/16 in.) long

TABLE 5.8   Cutting Conditions for the Automatic Tool Wear
Compensation Tests

Cutting speed	76.2 m/min (250 SFPM), 91.44 m/min (300 SFPM), and 106.68 m/min (350 SFPM)
Spindle speed	54, 65, and 76 rpm
Feed	0.254 mm/rev (0.010 in./rev)
Depth	0.508 mm (0.020 in.)
Coolant	Trimsol flood
Tool	883 carbide—0.794 mm (1/32 in.) (#TPMG 542) and 1.19 mm (3/64 in.) (#TPMG 543) nose radii
Inspection interval	29, 24, and 21 seconds

illustrated in Figure 5.21. As mentioned earlier, this setup
incorporated a very accurate micrometer graduated in tenths
(0.00254-mm or 0.0001-in. increments) to position the air nozzle relative to a fixed plate. Next, the micrometer was used
to good advantage to set up the table of air gap versus air
gauge output in the 1050T numerical control as described
previously.

This done, the air nozzle was moved onto the cutting tool
and the tests were started. Figure 5.10 has already shown
that flood coolant does not degrade air gauge accuracy. It is
also interesting to note that during the 2-day test series which
ensued only one nozzle was damaged. This failure was not due
to chip damage but instead to running an incorrectly adjusted
nozzle into the part. In this vein, it should be noted that the
tolerance on the inserts used was around ±0.0254 to 0.0508 mm
(0.001 to 0.002 in.). Thus, the position of the air nozzle relative to the tool tip varied from insert to insert. This made it
appropriate to adjust the air gauge with each insert change.
This was done by bringing the tool tip into contact with the
part and using a feeler gauge to adjust the air gap, after which
the nozzle securing screws were tightened. This procedure was
fast and repeatable.

The test procedure itself was not changed throughout the
course of the 2-day series of tests. It was used for both uncompensated and compensated cuts. In each case as a first

TABLE 5.9   Recorder Conditions for the Automatic Tool Wear
Compensation Tests

Chart speed	1 small division/second
Chart air gap scale	0.0254 mm (0.001 in.)/major division
Chart air gap range	0.0254 to 0.279 mm (0.001 to 0.011 in.)

step a straight cut was run using the cutting conditions listed
in Table 5.8. Air gauge data were recorded during the run.
The chart recorder was run at all times according to the conditions given in Table 5.9.

Following each cutting pass, an inspection pass was run to
inspect the profile generated by the cut. This was done by
retracting the tool a couple of thousandths of an inch from the
workpiece surface by MDI and rerunning the part program.
The amount of retraction used was such as not to exceed the
range of the air gauge and not to touch the part during the
inspection pass. The chart recorder was run during the inspection pass according to the conditions of Table 5.9, so it was
possible to align the recordings of the cutting and inspection
passes for direct comparison.

With the test procedure thus established, several uncompensated and wear-compensated tests were made using the three
cutting speeds and two insert types listed in Table 5.8. The
results of the machining trials follow.

Test Results

The 1/32 Inch Nose Radius Tool (TPMG 542)

As a first step in the test program the behavior of the machine
sensor, and recording system was evaluated by performing an
uncompensated cut on the titanium workpiece. The pertinent
conditions for the run are given in Tables 5.7 through 5.9. A
cutting speed of 76.2 m/min (250 SFPM) was used. The air
gap was recorded over the length of the cut. It will be remembered that the air gauge output during a cut indicates the tool
nose wear. Thus the result was a continuous record of tool
wear during the run.

Next, with the air gauge recorder running, the tool was
pulled out of contact with the work and the part program

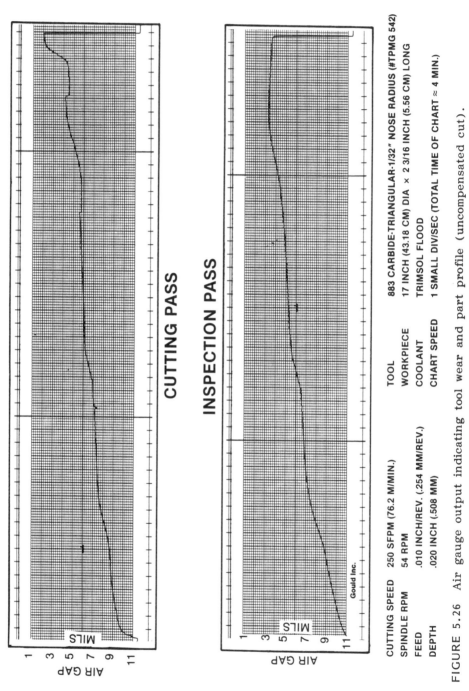

CUTTING PASS

INSPECTION PASS

TOOL	883 CARBIDE-TRIANGULAR-1/32" NOSE RADIUS (#TPMG 542)
WORKPIECE	17 INCH (43.18 CM) DIA × 2 3/16 INCH (5.56 CM) LONG
COOLANT	TRIMSOL FLOOD
CHART SPEED	1 SMALL DIV/SEC (TOTAL TIME OF CHART ≈ 4 MIN.)

CUTTING SPEED	250 SFPM (76.2 M/MIN.)
SPINDLE RPM	54 RPM
FEED	.010 INCH/REV. (.254 MM/REV.)
DEPTH	.020 INCH (.508 MM)

Gould Inc.

FIGURE 5.26 Air gauge output indicating tool wear and part profile (uncompensated cut).

180

(straight cut) was rerun. This inspection pass revealed the part contour generated by the cutting pass. Figure 5.26 shows the result. The top plot of this figure shows the air gauge output during the cut, while the bottom shows the results of the inspection pass. The time duration of the run was, according to the chart, about 4 minutes. During these 4 minutes, after some rapid initial nose wear, the tool wore about 0.127 mm (0.005 in.). Note that the bump at the end of the plot was due to the tool running off the end of the workpiece. With the resultant loss of cutting force, the tool sprung toward the work. The nozzle, since it trailed the tool tip (see Figure 5.8) recorded this springback as a 0.0508 mm (0.002 in.) reduction in the air gap. This effect, i.e., tool deflection, will be seen to be important in later tests, where a larger nose radius tool with increased cutting forces was used. At any rate, the sizable nose wear of 0.127 mm (0.005 in.) occurred during 4 minutes of cutting. The wear was not steady but accelerated and decelerated in a random way during the cut.

The question is, what effect did this wear have on the ultimate part profile? As would be expected, since a low tip radius tool with low cutting forces was used, the air gap profile of the inspection pass was nearly identical to the wear signal generated during the cut. This is true because as the tool wore during the cutting pass the air gap grew smaller and the finished part diameter grew larger. Thus, during the inspection pass as the air gauge proceeded along the part, the larger part diameter associated with increased tool wear resulted in a steadily decreasing air gap. Of course, the springback bump at the end of the cutting pass was not present in the inspection pass because there was no actual bump on the part. Next, consider wear compensation. The sensor arrangement used is important, so reference should again be made to Figure 5.8. It will be noted from the figure that the rear of the air nozzle lagged the cutting point by about 3.048 + 2.032 (nozzle diameter) = 5.08 mm (0.120 + 0.1080 = 0.200 in.). The significance of this delay was discussed in detail in the software section and it will be remembered that it was necessary to put a time delay in the numerical control compensation software to ensure that the nozzle would be fully over the surface generated by the previous compensation before another compensation step could be made. This delay fixed the inspection interval. At the feed used, the interval amounted to 29 seconds when the spindle speed was 54 rpm [76.2 m/min (250 SFPM) cutting speed]. For 91.44

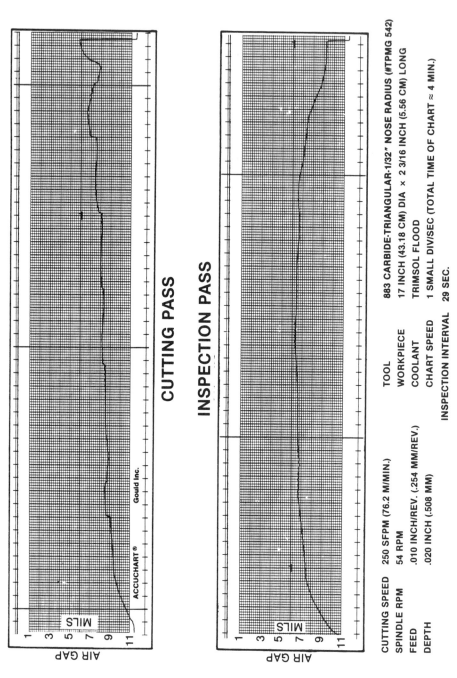

CUTTING PASS

INSPECTION PASS

CUTTING SPEED	250 SFPM (76.2 M/MIN.)	TOOL	883 CARBIDE-TRIANGULAR-1/32″ NOSE RADIUS (#TPMG 542)
SPINDLE RPM	54 RPM	WORKPIECE	17 INCH (43.18 CM) DIA × 2 3/16 INCH (5.56 CM) LONG
FEED	.010 INCH/REV. (.254 MM/REV.)	COOLANT	TRIMSOL FLOOD
DEPTH	.020 INCH (.508 MM)	CHART SPEED	1 SMALL DIV/SEC (TOTAL TIME OF CHART ≈ 4 MIN.)
		INSPECTION INTERVAL	29 SEC.

FIGURE 5.27 Air gauge output using tool wear, wear compensation steps, and part profile (compensated cut).

and 106.68 m/min (300 and 350 SFPM) cutting speeds the interval was 24 and 21 seconds, respectively. Therefore, 29 seconds after the cut was started the initial "zero" or no-wear air gap was placed in the 1050T's memory. At subsequent 29-second intervals, the air gap was examined and compared with its previous value. If it decreased, indicating wear, the amount of the decrease in tenths of mils was sent to the axis control board, which made a step in-feed of the x-axis tool position to compensate for wear.

Using this system, a compensated cut of the titanium part was made while recording the air gauge output. The test conditions were identical to those for the uncompensated cut. A fresh insert tip was used. Figure 5.27 shows the air gap plots for the compensated cut and its associated inspection pass.

Compensations for wear on the cutting pass recording are shown as abrupt closures of the air gap corresponding to the step changes in the x-axis position. Following these step closures, the gap generally started to reopen after about 10 seconds when the air gauge started to inspect the surface generated following the change. Thus, after each step closure there was something of a reopening, but the general trend of the curve was the same as that of any other wear curve. In fact, the average curve for a compensated run represents tool wear just as it does for an uncompensated cut.

Following the compensated cut, an inspection pass was again performed with the tool and work out of contact. The inspection pass is shown as the bottom curve of Figure 5.27. With the exception of the beginning and ending portions, the curve is essentially flat, indicating that the part's geometry had been compensated automatically for tool wear to within about 0.0254 mm (0.001 in.).

Another test, identical to that shown in Figure 5.27, was run with a new cutting tip. Figure 5.28 shows the results. After the initial compensation was made about 58 seconds into the run, the tolerance on the part profile was again held to within 0.0254 mm (0.001 in.). The tool wore much faster during this run and consequently the correction steps were much larger. Here, after initial break-in the tool wore 0.016 mm (0.004 in.) compared with 0.0508 mm (0.002 in.) for its companion run shown by Figure 5.27. In spite of this sizable

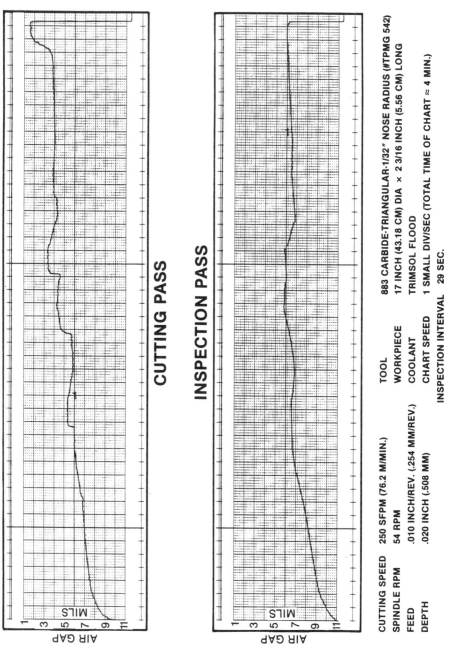

Figure 5.28 Air gauge output indicating tool wear and part profile (compensated cut).

difference in tool wear, which, no doubt, arose from variations in the tool and workpiece material etc., the part profiles were held to within 0.0254 mm (0.001 in.).

By comparing the uncompensated inspection curve of Figure 5.26 with the compensated profiles of Figures 5.27 and 5.28, the effects of tool wear compensation on part geometry can be clearly seen. The curves demonstrate the technical feasibility of continuous closed-loop machining.

In the interest of clarity, some other points should be made regarding Figures 5.26 through 5.28. For example, if the compensations were put in as step changes, why didn't they appear as steps in the part profile during inspection? There were, in fact, steps in the part profile, but the slow passage of the 2.032 mm (0.080 in.) diameter nozzle over them tended to wash out any abrupt changes in the air gap output and thus only the average shape is shown by the inspection pass.

Figures 5.26 through 5.28 clearly indicate the need for automation: even though everything was the same for the three cuts represented, the total tool wear was 0.127, 0.0508, and 0.1016 mm (0.005, 0.002, and 0.004 in.) for Figures 5.26, 5.27, and 5.28, respectively. But, even though the amounts of wear varied widely for the three cuts, the force at the end of each cut as indicated by the springback was about the same, the springback being about 0.0508 mm (0.002 in.) in each case.

After several tests were run under the conditions used for Figures 5.26 through 5.28, the supply of TPMG 542 tools [0.794 mm (1/32 in.) nose radius] was exhausted. TPMG 543 inserts were available, so the testing was continued with them. The TPMG 543 is identical to the TPMG 542 in all respects, except that it has a 1.191 mm (3/64 in.) instead of a 0.794 mm (1/32 in.) nose radius. The increased nose radius led to higher normal cutting forces. The machine's behavior was vastly changed by the higher forces.

The 3/64 Inch Nose Radius Tool (TPMG 543)

Several tests were run using new TPMG 543 inserts. To increase the wear rate, a somewhat higher cutting speed of 91.4 m/min (300 SFPM) was used for these tests. Figure 5.29 shows the results of the first test run under the modified conditions. The test was identical in all respects to those described

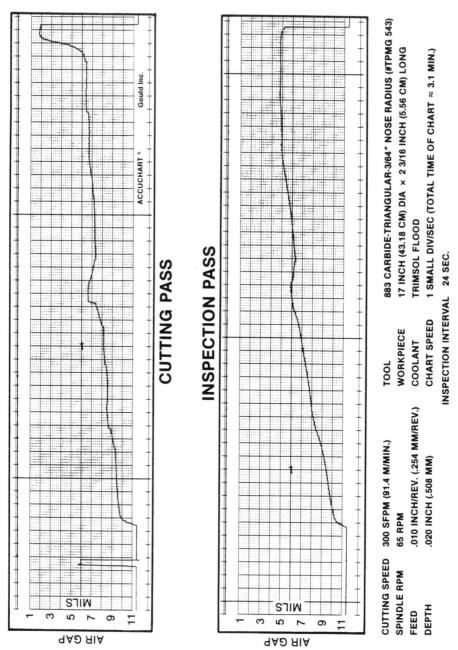

FIGURE 5.29 Compensated cut using large nose radius insert.

previously, except for the tool and speed already mentioned and except that a time delay of 24 instead of 29 seconds was used.

The cutting pass in Figure 5.29 looks very similar to those in Figures 5.26 and 5.28, and several compensation steps can be observed. One notable difference between the curves is the springback at the end of the cut. Because of the higher normal force, the springback was considerably larger for the larger nose radius insert.

But consider the inspection pass for Figure 5.29 as compared with Figures 5.27 and 5.28. It is clear from Figure 5.29 that the part profile was not well compensated for wear in the case of the larger nose radius tool. Two reasons can be given for this behavior. One is related to the cutting force and the other is related to the machine's response to compensation steps under elevated cutting forces.

To get a feeling for the second of these causes, another test, identical in all respects to the one illustrated in Figure 5.29, was run. For this test, however, a record was kept of the x-axis resolver of the Gray VTL and an independent Sony scale that read the x-axis tool carriage position in tenths of mils. Figure 5.30 shows the chart recorder output for this run and its inspection pass. Comparison of Figures 5.29 and 5.30 shows that, in general, the runs were nearly identical in all respects. Time markers, however, are included in Figure 5.30 to indicate the 24-second sampling intervals for the wear compensation software. It will be remembered that this interval is needed to compensate the lag of the nozzle behind the tool tip.

As stated, a log of the Gray VTL's x-axis resolver and Sony scale was kept. That log is shown in Table 5.10. Now, compare, step by step, Table 5.10 with the cutting pass of Figure 5.30. As the starting point note on Figure 5.30 the initialization air gap for the 1050T software at 24 seconds from the start of cutting. It was 0.264 mm (0.0104 in.). Looking at the first compensation interval at 48 seconds, the air gap was 0.254 mm (0.0100 in.), or a change of 0.0102 mm (0.0004 in.). According to Table 5.10, this agrees closely with the resolver change at 48 seconds, which means that the 1050T software responded correctly to the change caused by tool wear and stepped the ball screw accordingly.

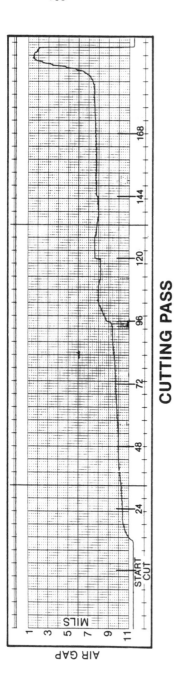

CUTTING PASS

INSPECTION PASS

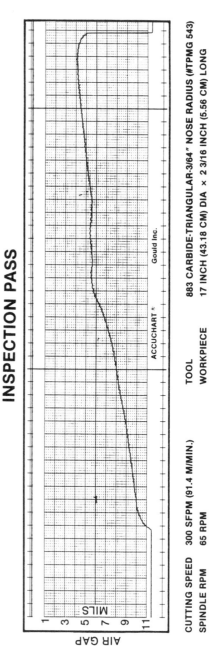

CUTTING SPEED	300 SFPM (91.4 M/MIN.)	TOOL	883 CARBIDE-TRIANGULAR-3/64" NOSE RADIUS (#TPMG 543)
SPINDLE RPM	65 RPM	WORKPIECE	17 INCH (43.18 CM) DIA × 2 3/16 INCH (5.56 CM) LONG
FEED	.010 INCH/REV. (.254 MM/REV.)	COOLANT	TRIMSOL FLOOD
DEPTH	.020 INCH (.508 MM)	CHART SPEED	1 SMALL DIV/SEC (TOTAL TIME OF CHART ≈ 3.1 MIN.)

INSPECTION INTERVAL   24 SEC.

FIGURE 5.30   Compensated cut using large nose radius insert.

TABLE 5.10 X-Axis Resolver and Sony Scale for the Test Illustrated by Figure 5.30

Time (s)	Resolver		Resolver change		Sony scale		Sony change	
	mm	in.	mm	in.	mm	in.	mm	in.
0	—	—	—	—	—	—	—	—
24	23.4061	0.9215	Init. zero	Init. zero	0.0051	0.0002	0	0
48	23.4137	0.9218	0.0076	0.0003	0.0076	0.0003	0.0025	0.0001
72	23.4213	0.9221	0.0076	0.0003	0.0178	0.0007	0.0102	0.0004
96	23.4366	0.9227	0.0153	0.0006	0.0533	0.0021	0.0356	0.0014
120	23.4620	0.9237	0.0254	0.001	0.0864	0.0034	0.0330	0.0013
144	23.4721	0.9241	0.0101	0.0004	0.0940	0.0037	0.0076	0.0003
168	23.4772	0.9243	0.0051	0.0002	0.0991	0.0039	0.0051	0.0002
192	23.4772	0.9243	0	0	0.0991	0.0039	0	0
			0.0711	0.0028			0.0940	0.0037

However, although the resolver stepped 0.0076 mm (0.0003 in.), the x-axis slide did not respond. This is evidenced by the fact that Figure 5.30 shows no step gap closure at the 48-second point like those of Figures 5.27 and 5.28. At the same time, the Sony scale indicated a 0.0025 mm (0.0001 in.) change.

Moving on step by step, look now at the 72-second interval. Figure 5.30 indicates a wear of 0.0076 mm (0.0003 in.) between 48 and 72 seconds, which was exactly the change indicated by the resolver. The Sony scale indicated a 0.0102 mm (0.0004 in.) change, and again no actual compensation is indicated by a step air gap closure in Figure 5.30.

During the next interval the wear rate increased and according to the figure the change from 72 to 96 seconds was 0.0127 mm (0.0005 in.). The resolver again responded, this time with a change of 0.0152 mm (0.0006 in.), and this time the x-axis slide moved abruptly as evidenced by the 0.0102 mm (0.0004 in.) step in the air gap of Figure 5.30. To this point the resolver had changed 0.0305 mm (0.0012 in.), but the actual slide had responded only 0.0102 mm (0.0004 in.). The Sony scale, however, indicated a change of 0.0356 mm (0.0014 in.).

At this point it is clear that the 1050T control was responding very reliably and converting air gauge wear indications into position commands, to which the lead screw drive was responding as indicated by the resolver. But it is also clear that the x-axis slide was not responding adequately to these changes. If one carries this step-by-step investigation to the end of the run, it is clear from Table 5.10 that the resolver indicated a 0.0711 mm (0.0028 in.) infeed of the x-axis slide, which is in exact agreement with the total wear indicated by the air gauge. The actual summed compensation jumps of the x-axis slide, however, amount to only 0.0304 mm (0.0012 in.). At the same time the Sony scale indicated that a 0.0940 mm (0.0037 in.) correction had been made.

The only conclusion one can draw from these results is that the x-axis slide pivoted or executed some motion such that the compensation was amplified at the Sony measurement point and retarded at the cutting point. This effect must be attributed to the increased force associated with the large nose radius tool used for these tests, because it did not occur during the tests of the small nose radius tool. For example, according to Figure 5.27, which represents small nose radius conditions, the total

tip wear was 0.0533 mm (0.0021 in.) and the sum of the correction steps was also 0.533 mm (0.0021 in.). Similarly, in
Figure 5.28 the total wear was about 0.0889 mm (0.0035 in.)
and the total of the compensating steps was 0.0787 mm (0.0031
in.). In these two runs the normal tool forces were apparently
low enough that they did not cause the slides to twist or stick.
It is unfortunate that the resolver and Sony outputs were not
recorded for these runs.

Referring back to Figure 5.30, remember that a correction
of 0.0711 mm (0.0028 in.) was input through the resolver, but
only 0.0305 mm (0.0012 in.) can be accounted for by the compensation steps. Therefore a profile error of 0.0406 mm (0.0016
in.) can be attributed to inadequate compensation. However,
according to the inspection pass for this run, the profile error
was about 0.1016 mm (0.004 in.) after compensation was initiated
at 48 seconds into the cut.

This effect can be attributed to the second error source
mentioned, namely the buildup of force with wear. As the tool
wore, the normal force increased, forcing the tool out of the
cut. This resulted in a larger than expected part diameter,
which on the inspection pass showed up as a decreasing air gap.
This did not happen for the tests of the small radius tool because the wear was adequately compensated.

In summary, the error due to the effect of wear on force
is not as serious as the sticking slide error because it can
easily be factored out by modifying the air gauge voltage/air
gap table in the numerical control. The twisting, sticking slide
problem, on the other hand, is stochastic in nature and is
therefore hard to compensate.

To demonstrate the effect on part profile of the very strong
wear-induced force change, an uncompensated cut was run using
a fresh large nose radius tool (TPMG 543) and a still higher cutting speed of 106.68 m/min (350 SFPM). The results of this
test are shown in Figure 5.31.

At the start of the run the air gap was around 0.2286 mm
(0.009 in.) and at its completion the gap had closed to 0.0914
mm (0.0036 in.), indicating a total tool tip wear of 0.1372 mm
(0.0054 in.) for the cut. The inspection pass, on the other
hand, started at 0.2667 mm (0.0105 in.) and the wear-related
taper of the titanium spool was so large that about three-fourths

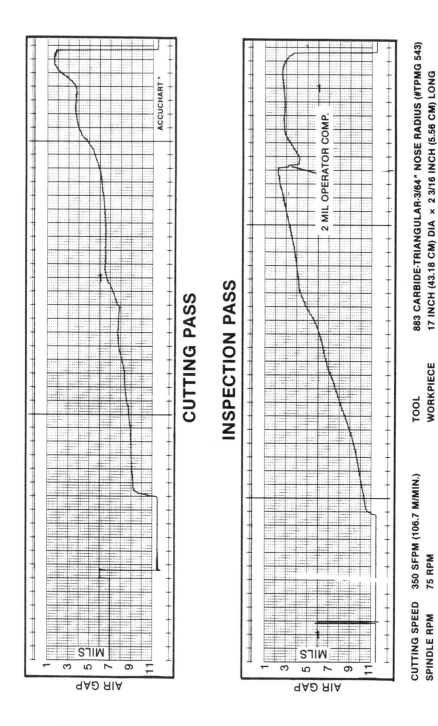

CUTTING PASS

INSPECTION PASS

2 MIL OPERATOR COMP.

ACCUCHART®

CUTTING SPEED	350 SFPM (106.7 M/MIN.)	TOOL	883 CARBIDE-TRIANGULAR-3/64" NOSE RADIUS (#TPMG 543)
SPINDLE RPM	75 RPM	WORKPIECE	17 INCH (43.18 CM) DIA × 2 3/16 INCH (5.56 CM) LONG
FEED	.010 INCH/REV. (.254 MM/REV.)	COOLANT	TRIMSOL FLOOD
DEPTH	.020 INCH (.508 MM)	CHART SPEED	1 SMALL DIV/SEC (TOTAL TIME OF CHART ≈ 2.8 MIN.)

FIGURE 5.31 Uncompensated cut using a large nose radius insert.

192

TABLE 5.11 X-Axis Resolver and Sony Scale for the Test Illustrated by Figure 5.31

Time (s)	Resolver		Resolver change		Sony scale		Sony change	
	mm	in.	mm	in.	mm	in.	mm	in.
0	—	—	—	—	—	—	—	—
21	24.2011	0.9528	Init. zero	Init. zero	0	0	Init. zero	0
42	24.2087	0.9531	0.0076	0.0003	0.0102	0.0004	0.0102	0.0004
63	24.2418	0.9544	0.0331	0.0013	0.0406	0.0016	0.0304	0.0012
84	24.2418	0.9544	0	0	0.0406	0.0016	0	0
105	24.2545	0.9549	0.0127	0.0005	0.0533	0.0021	0.0127	0.0005
126	24.3129	0.9572	0.0584	0.0023	0.1524	0.0060	0.0991	0.0039
147	24.3357	0.9581	0.0228	0.0009	0.1701	0.0067	0.0178	0.0007
			0.1346	0.0053			0.1702	0.0067

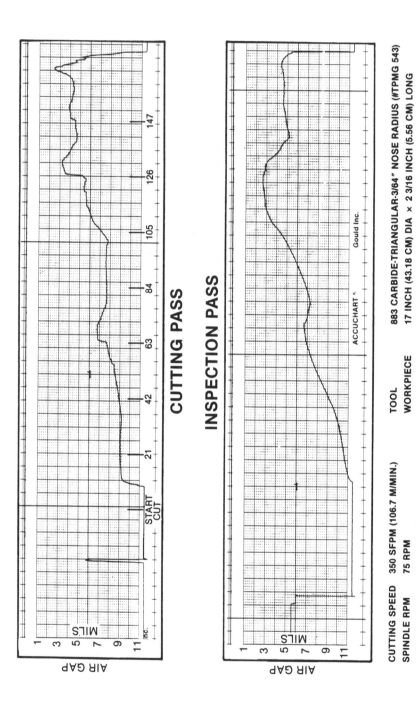

FIGURE 5.32 Compensated cut using large nose radius insert.

194

of the way through the pass the operator had to manually input
a 0.0508 mm (0.002 in.) withdrawal of the tool to prevent it
from running into the work. The final gap after the operator
compensation was 0.0660 mm (0.0026 in.), so that the diameter
varied 0.2515 mm (0.0099 in.) over the length of the part. Of
this total, 0.1372 mm (0.0054 in.) can be attributed to wear
itself. The remaining 0.1143 mm (0.0045 in.) was due to the
force increase associated with tool wear during this uncompensated cut.

To complete this section consider finally a wear-compensated
cut run under conditions identical to Figure 5.31 with a fresh
TPMG 543 (3/64 in. nose radius) insert. Data was taken from
the x-axis resolver and the Sony scale during the run after
each 21-second compensation interval. Table 5.11 shows the
resolver and Sony outputs for the cut and Figure 5.32 shows
the air gauge output for the cutting and inspection passes.

If one now goes step by step through the table and Figure
5.32, as done with Table 5.10 and Figure 5.30, similar behavior
will be observed. For example, at the 42-second increment, the
air gauge called for a 0.0076 mm (0.0003 in.) infeed (see Figure
5.32), which the resolver correctly indicated. The slide, however, did not increment the tool this amount, but the Sony
scale picked up a 0.0102 mm (0.0004 in.) change. Similarly,
the system responded as it had done previously for the reat
of the run. Sometimes the slide responded; at other times it
didn't. One can be satisfied of this by stepping through the
table and figure simultaneously.

For the purpose of this description, however, consider instead the end result. The total wear of the tool in the compensated zone between 21 seconds and the start of the springback
peak was, according to Figure 5.32, 0.1245 mm (0.0049 in.)
(9.1 - 4.2 = 4.9 mils). It was very close to the companion uncompensated cut (0.1372 mm or 0.0054 in.) illustrated by Figure
5.31. Because of sticking, the sum of all the actual compensations was 0.0737 mm (0.0029 in.), so a direct wear-related profile error of 0.0508 mm (0.002 in.) (4.9 - 2.9 = 2.0 mils) resulted. On top of this, a 0.1143 mm (0.0045 in.) error can be
attributed to the wear-induced increase in cutting force. This
force magnification effect on the part's geometry is clearly
shown in Figure 5.32.

By combining the two errors, one would expect a total be-
ginning-to-end part error for the inspection pass of Figure 5.32
to be 0.1651 mm (0.0065 in.) (2.0 + 4.5 = 6.5 mils). The inspection pass of the figure is in exact agreement with this reasoning,
indicating that the air gauge instrumentation had accounted for
the complex machine behavior.

CONCLUSION

A new approach to in-process inspection and tool condition compensation was described in this chapter. It was shown that a
tool-mounted air gauge displacement sensor can accurately measure tool wear on the fly and a control loop closed through the
numerical control can automatically correct part shape for wear.

It was also shown that under conditions where the system
doesn't adequately correct for wear, the air gauge can explain
the error sources. The method was shown to be powerful in
that it can measure faulty machine tool behavior (sticking slides
and bearing runout) and behavior related to machine/part deflection under the buildup of force with tool wear.

Since the error-causing behavior of the machining process
can be understood by this approach, it should work well for
on-line adaptive control.

Next steps in the work should be directed toward applying
the methods described here to contoured surfaces and finally
expanding the use of dynamic in-process inspection to the automatic inspect/rework cycles as shown previously in Figure 5.2.

REFERENCES

1. D. G. Flom, "Manufacturing Initiative for Advanced Metal
    Removal." Air Force Contract F33615-80-c-5057 (AFWAL-
    TR-85-4044), May 1985.
2. Ibid. Vol. 4, "Advanced Metal Removal Technology," S. D.
    Murphy 4-2, "Establishment of an In-Process Inspection
    System."
3. W. J. Blaiklock, "Air Gaging, Then and Now," Edmunds
    Manufacturing Company, Farmington, CT 06032.
4. R. A. Thompson, GE Internal Report. Contact the author
    for more information.

6

In-Process Inspection System Using Tool-Touch Auditron

WILLIAM S. McKNIGHT* / General Electric Company, Cincinnati, Ohio

This chapter describes the development of an in-process inspection system which is currently being used in production to close the loop around a turning operation in the fabrication of rotating components for gas turbine engines, primarily compressor and turbine disks (spools). The system's in-process sensor is a Tool Touch Auditron (TTA). The TTA is a contacting, indirect device which plays an integral role as a static sensor in a closed-loop machining (CLM) system which was retrofitted onto a turret lathe.

The work described here was performed under contract to the United States Air Force (Ref. 1). The overall objective of the work effort was to increase the productivity of NC turning operations by reducing the noncutting time associated with tool setting and manual in-process inspection operations. This was accomplished through the development of the TTA sensor, which was initially tested for its feasibility as an in-process sensor and subsequently ruggedized for production implementation. An economic model was also developed and was later corroborated by production implementation which resulted in both labor and scrap savings.

---

*Present affiliation: Belcan Engineering Services, Inc., Cincinnati, Ohio.

TOOL TOUCH AUDITRON CONFIGURATION

Figure 6.1 shows a representation of the CLM system that was developed. The TTA is mounted to the vertical axis of a vertical turret lathe (VTL) with its cutting tool serving as the contacting probe. The output signal from the acoustic emission transducer is connected to a computer, as are the signals from the position transducers (encoders) attached to the machine's slides. The TTA provides a signal on contacting the workpiece; the probe depends on the encoders to provide the positional coordinate measured relative to a fixed datum. Output from the computer is in turn connected to the NC controller, which provides the correction for the machining operation. The system determines deviation from the expected nominal dimensions and is capable of adjusting the programmed tool path to cancel the measured deviations for subsequent tool motions.

The TTA is configured as a static-type system; that is, measurements are made in between cutting operations. Since the transducer is an acoustic type, it requires the spindle of the lathe to be rotating while a measure is being made. The system was implemented on a Gray VTL with a GE 1050 CNC

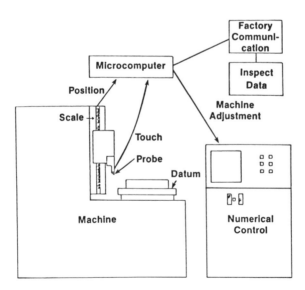

FIGURE 6.1 Closed-loop machining.

controller. An acoustical sensor was used with closed-loop machining software that were developed at GE. A magnetic scale system manufactured by SONY provided the positional information.

After the sensor was initially developed, it was modified to make it more suitable for the production environment. The modifications included more rugged packaging, improved signal processing, and better noise immunity. The performance of the in-process inspection system is highly dependent on the performance of the TTA sensor. The repeatability, reliability, and response time (in conjunction with the system response time) of this sensor provide, with few exceptions, the basis for the performance characteristics of the entire system. For this reason, the demonstration of this critical subsystem was considered to have the highest priority in the effort to develop the in-process system.

EVALUATION OF TOUCH LOCATION REPEATABILITY

A performance plan was put together to evaluate the touch location repeatability of the TTA. This plan consisted of a factorial experiment using a fixed configuration of the sensor on a Gray VTL in conjunction with a GE Mark Century 1050T CNC/CLM system using high-confidence statistical populations of data. Two alloy systems (nickel and titanium) were used in the evaluation to give a good representation for aerospace applications.

The sensor was evaluated in its ruggedized form using the printed circuit board design, as opposed to the earlier hand-wired versions. The acoustic emission transducer was a Dunegan/Endevco D750. The specified compressional wave frequency bandwidth for this transducer is 100 to 300 kHz with a sensitivity of 60 db at its resonant peak frequency. The same transducer had been used in earlier feasibility testing at GE with the Gray VTL and other machines for both touch sensing and closed-loop machining. The amplifier was the same one used in earlier designs. The unit consists primarily of an active high-pass filter with thresholding and analog-to-digital conversion to provide an appropriate signal for the CNC controller. The updated printed circuit board version of the sensor made provisions for wire wrap selectable filtering. The filter's high-pass cutoff frequency can be changed via a pad of two resistors and two capacitors. The amplifier gain is easily changed via a pad of four resistors. The amplifier accommodates both single-ended and differential acoustic emission sensors through wire wrap jumpers. The D750 transducer utilized provides a differential signal; thus the amplifier was configured accordingly for the evaluation.

For the evaluation, the amplifier was set to a gain of 200. This was sufficient to compensate for losses through the low-pass filter (approximate factor of 4 loss). The low-pass and cutoff frequencies were set at values used in earlier designs. These values were kept constant during the course of the experiments. The four comparator coarse threshold resistors were kept at previous values. Fine threshold adjustments of around 1.0 volt were made using the associated potentiometer as an evaluation variable. The baseline threshold was set at 0.075 volts, the peaks threshold at 0.500 volts, and the counts threshold at 6.0 volts.

Two digital discrimination threshold values are associated with the events and peaks parameters. These were adjusted through their associated binary switches as an evaluation variable. The counts threshold remained at a fixed setting determined at the time of setup of the Gray VTL. The logically "ANDed" result of touch "success" in each of the three channels determined whether or not a "touch" signal was passed on to the MC 1050 controller. The updated version of the sensor contained logic inversion circuitry to accommodate a "loss of acoustic vibration" signal, which was used to detect a tool breakage event. This feature was not evaluated during the experimental trials, however.

The combination machine tool and control was configured as shown schematically in Figure 6.2. It features the Gray vertical turret lathe as it was in GE's Manufacturing Technology Operating/Machinery Technology Laboratory (MTL/MTO) laboratory, which serves to provide all touch location measurement data via its encoders and the MC 1050T controller. The latest-version CLM software supports the data acquisition required for the experiments. A special process program tape was generated to supply appropriate machine motions and CLM functions. This program featured a series of subroutines which performed a population of 30 repetitions for each set of test conditions. Each touch of the sensor occurs along a "fresh

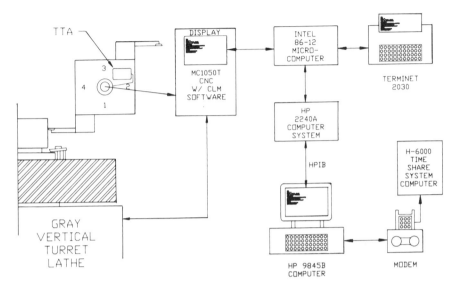

FIGURE 6.2   Schematic representation of machine tool and control.

track" on the part surface.   Incremental touch track positioning was included in the program tape.   Manual part-shim-tool
measurements were made to establish tool offsets which relate
surface location values on the program tape to actual part surface contact location in machine space.   No datum surfaces were
necessary to accomplish the subject evaluation.   The recommended tool insert material was carbide, which was appropriate
to the workpiece material.   The edge was worn (or preworn)
during reference surface-finishing cuts.   The objective was to
have the tool operate along its normal wear curve in a regime
where tool chipping would not normally occur.   Data acquisition
was accomplished automatically by CPC's Hewlett-Packard 9845B
computer system, combined with the companion 2240A measurement and control unit.   They were programmed to take raw
data from the MC 1050T display panel logic circuits through
parallel data lines through an Intel microcomputer.   The bit
patterns were interpreted as displayed characters, logged into
memory, and transferred to magnetic tape cassettes.   The values so logged were used for statistical analyses in the subject
evaluation.   False touches were noted by hand.

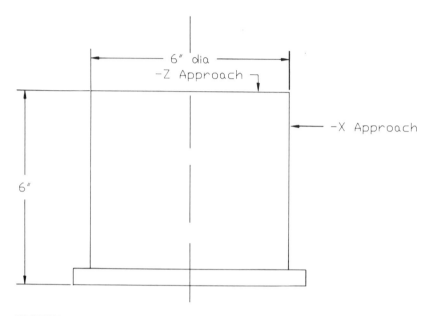

FIGURE 6.3   Right cylinder.

In order to simulate actual part geometry, i.e., a gas turbine disk, a true right, round cylinder was utilized as depicted
in Figure 6.3.   This shape is intended to represent part features that would be turned.   Radial runout was kept to within 0.0003 in., axial taper (conics) was within 0.003 in./ft, and
face taper was also within 0.003 in./ft.   The cylinder provided
an external (outer diameter) surface (approach in the -X direction) capable of providing at least 15 increments of axial position for a set of touch tracks.   The cylinder provided a top
face (approach in the -Z direction) for a similar set of touch
track increments.   All touch track zones on the cylinder were
evenly coated with blue marking dye before touch location measurements.   The cylinder was manually measured whenever a
new surface was generated in order to quantify the overall dimensions, runout, and conics.

The touch location repeatability evaluation was based on
numbers produced by the MC 1050T controller.   These numbers
combine the ability of the TTA to detect and discriminate actual
touch and the ability of the CNC software and hardware to capture machine tool resolver positions in a variable time span after

being notified by the TTA that a touch has occurred. The response capability of the TTA was not isolated in this study.
The subject evaluation characterized the combined responses
for a typical rotating machine tool. Such analysis is most
practical and useful for CLM applications planning.

The machine tool with its CNC controller served as the part
dimension qualifier for touch measurements and calculations. The
governing relationships are described as follows:

Actual dimension = tape dimension - tool offsets

Corrected touch location = resolver reading + compensations

Deviation = actual dimension - corrected touch location

This deviation value was displayed by the CLM software and
captured by the Intel/Hewlett-Packard computer system.

The following variables were selected to give a comprehensive, practical evaluation for the experimentation necessary to
establish the performance of the sensor that appropriately used
limited resources to collect and process the resultant data:

Data repeat set size: 15
Alloys: two [Inconel (nickel alloy) and titanium]
Turret positions: four (1, 2, 3, 4)
Surface velocities: six (83 ± 7 SFM for Inconel and 198 ± 28
  for titanium)
Approach rates: two (1 and 5 IPM)
Analog thresholds: two (events)
Digital thresholds: six (peaks and events)
Total data sets: 960
Total touch measurements: 14,400
Assumed touch measurement cycle time: 20 seconds
Data set acquisition time: 10 minutes
Minimum 7-hour days for all data: 14 days

The data were acquired with the data repeat increments as
the fastest changing variable, the digital threshold as the next
fastest, and so on up the list until the slowest changing variable, the alloy, was completed.

A test procedure was established to conduct the experiments.
First, the Inconel cylinder was mounted to the lathe and subsequently turned to create the true right cylinder with its four
reference surfaces. The experiments were then run until the

data combinations were exhausted. When a surface measurement
zone was used up, the surface was photographed for a perman-
ent visual record, and then the surface was remachined to create a fresh measurement area. When all tests on the Inconel
cylinder were completed, the titanium cylinder was mounted and
the procedure was repeated with the second alloy. After all
data collection, the information was analyzed to determine statistical significance and to identify major and minor interactions.
Repeatability and variation about the mean were computed and
the data combinations were rated according to their ability to
yield the best touch deviation repeatability. The combinations
yielding the best sensor performance were recorded for future
use in a production environment.

Three additional tests were performed with the sensor using
the optimum set of operating conditions in order to confirm the
repeatability and reliability of the TTA sensor. Data from a set
of 1000 touches were collected and analyzed for tests 1 and 2 on
the Inconel 718 and 6-4 titanium cylinders. The procedure and
equipment were the same as in previous tests except that several
changes were made to the TTA and its interface which increased
the signal-to-noise ratio. These changes are described later in
this chapter. The third test was a 300-touch repeatability evaluation of the sensor with a Kyon insert using the Inconel cylinder. This last test was performed because of a direct production application opportunity.

The data analysis performed after testing quantified the
TTA performance. An analysis of variance provided F ratios
for each combination of factors or parameters. This value is
a measure of the significance or degree of effect of a particular parameter on the overall repeatability (variance) and average touch depth of the TTA sensor. The F ratios for all parameters evaluated on the titanium and Iconel cylinders are
ranked in order of significance and shown in Tables 6.1 and
6.2, respectively. The adjustable thresholds in the signal processing circuitry of the sensor appear to have less influence on
the system's repeatability in the titanium test than in the Inconel
test. Figure 6.4 shows the excellent repeatability of the sensor
at appropriate operating conditions with little dependence on surface velocity. Figure 6.5 contains a less repeatable performance
that was observed under similar conditions. The data for this
set (and several others) were found to be randomly contaminated
with spurious values that had little effect on the average depth

TABLE 6.1 TTA Performance Evaluation on Titanium: Statistical Analysis of Variance Ranking by F Ratio

REPEATABILITY

Parameters	F ratio
Approach feed rate[a]	22.7008
Peak D and events D	7.5354
Surface velocity[a]	6.8774
Events D threshold	6.8485
Events A and peaks D and turret position	5.8007
Events A and peaks D threshold	5.3833
Events D and turret position	4.8467
Events D and surface velocity	4.1482
Peaks D threshold	4.0748
Events analog and peaks D and surfaces	3.5959

AVERAGE TOUCH DEPTH

Parameter	F ratio
Turret position[a]	44.1058
Peaks D and events D	38.3084
Events D threshold	34.8729
Events A and peaks D and turret	33.3952
Peaks D and events A	15.9867
Surface velocity[a]	15.0894
Peaks D and turret position	9.7394
Events A and events D	8.9384
Events A and events D and turret	7.7083
Peaks D	6.4775
Events A	6.3874
Events A and peaks D and events D	5.7439
Events D and turret position	4.4105
Events D and surface velocity	4.2948
Events A and peaks D and surface	3.6288
Approach feed rate[a]	3.4032

[a]Machine-dependent parameters.

TABLE 6.2 TTA Performance Evaluation on Iconel: Statistical Analysis of Variance Ranking by F Ratio

REPEATABILITY

Parameters	F ratio
Peaks D threshold	145.47
Surface velocity[a]	69.60
Events A threshold	48.47
Approach feed rate[a]	43.27
Turret position[a]	17.29
Events D threshold	17.17
Peaks D and surface	16.16
Peaks D and events D	14.90
Events A and turret	14.73
Events A and peaks D	12.79
Events D and turret	7.41
Events A and surface	5.99
Peaks D and approach	5.26
Peaks D and turret	4.79
Events D and surface	3.25
Surface and approach	2.04
Events A and events D	1.91
Turret and surface	1.44

AVERAGE TOUCH DEPTH

Parameters	F ratio
Peaks D threshold	129.12
Surface velocity[a]	125.58
Turret position[a]	79.10
Peaks D and events D	75.07
Events D threshold	72.19
Events A threshold	51.38
Events D and turret	39.51
Approach feed rate[a]	37.76
Events A and turret	23.20
Peaks D and events D and turret	20.48
Events A and peaks D and turret	14.45
Peaks D and turret	13.54
Peaks D and events D and surface	10.56
Events D and surface	8.48
Peaks D and approach	7.34
Events A and events D and turret	6.46
Events A and events D	4.68
Events A and surface	4.41
Peaks D and surface	3.64

[a]Machine-dependent parameters.

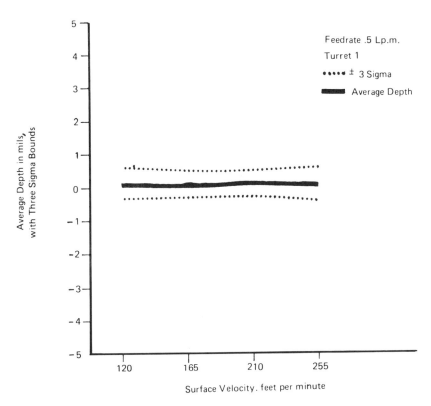

FIGURE 6.4   Typical TTA repeatability.

but caused the indicated loss of repeatability. It was determined
that this behavior was typical with incipient transducer failure.
Replacement of the transducer corrected the problem.

With the optimum operating conditions determined from the
previous tests, the TTA and in-process inspection systems were
programmed to make 1075 touches on the surface of the titanium
cylinder. The results of this more extensive test compared favorably with previously reported repeatability results. No fail-
ures or spurious or unexpected values were observed. The
average touch depth into the surface made over the 1000-point
sample was 0.00025 in. (0.0064 mm) with a repeatability (3 sigma) of 0.00045 in. (0.0114 mm) for 99.5% of all expected values
at a 95% confidence level. Since this test extended over several

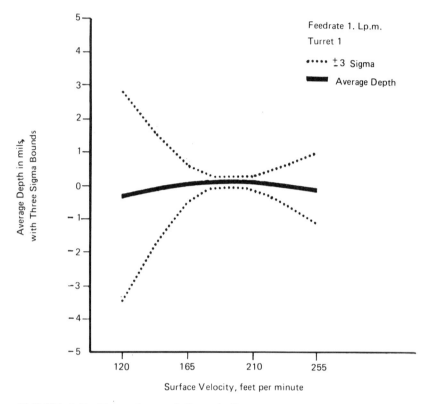

FIGURE 6.5 Transducer failure influence on repeatability.

days, this repeatability includes variances associated with start-up and day-to-day operation.

Before the 300 final touches were made, the cylinder was coated with DyKem so that touch impressions could easily be located and inspected. Figure 6.6 shows one of these touch impressions at 15× magnification with part of the DyKem wiped away and the machined surface texture exposed. The horizontal lines from successive tool paths are readily visible. The impression made by the tool via the TTA sensor is a narrow streak extending through the cleaned area. A second view of this is shown in Figure 6.7, this time at a magnification of 40×, where it is evident that the depth of the touch mark is smaller than the original machined texture. This was verified with a

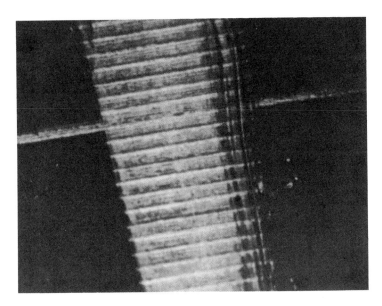

FIGURE 6.6   TTA touch impression (15×).

profilometer, despite the fact that there is really insufficient resolution at these small magnitudes.

Another test with 1075 touches similar to the previous test was performed on the Inconel cylinder. The results were expectedly similar to those from the titianium test. The average touch depth was 0.0003 in. (0.0076 mm) with a repeatability (3 sigma) of 0.00057 in. (0.0145 mm). One interesting difference between the two materials was that the average depth was indicated as being into the surface (a negative dimension) in the case of titanium but away from the surface (a positive dimention) in the case of the nickel alloy cylinder. This suggests that the TTA may have sensed the surface before it actually touched it! This can be explained by reviewing Figure 6.7, where the touch impression was on the cusp between two adjacent tool path grooves from the original machined surface. A touch depth of zero would coincide with a touch at the bottom of the tool path groove. A negative (into the workpiece) touch

FIGURE 6.7   TTA touch impression (40×).

depth would cause a deeper tool path groove, whereas a positive (away from the workpiece) touch depth would not be as
deep as the groove depth caused by the tool itself, as illustrated precisely in Figure 6.7.

Because of the application opportunity of the CLM using
KYON inserts, an additional 300-touch test was made.   The 300
data points were collected using these inserts on the Inconel
cylinder.   The average touch depth was observed to be 0.00015
in. (0.0038 mm) into the workpiece with a 3 sigma repeatability
of 0.00075 in. (0.019 mm).   This agreed favorably with the
other results.

TRANSDUCER EVALUATION

Since the transducer used in the TTA, a Dunegan-Endevco type
750, was quite hard to obtain, it was felt that other transducers

TABLE 6.2   Evaluation Results of TTA Replacement Transducers

Manufacturer	Model	Frequency	Light scratch response (mV)	Light tap response (mV)
Endevco	D750	200	75	150
B & K	4375	92	75	200
Physical Acoustics	R-15	150	200	300
Physical Acoustics	T-30	300	100	200

should be evaluated for their performance in the sensor.   Four
alternative transducers were procured for evaluation.   The
Physical Acoustics Corporation, Princeton, N.J., supplied three
transducers for the TTA design.   The model R-15 transducer
is an omnidirectional, floating, differential design with excellent
sensitivity (70 db) at a resonant frequency of 150 kHz.   Model
R-15-I is the same as the previous type except that it has an
integral preamplifier which provides an additonal gain of 40 db.
The third was a T-30 model which is tuned to a frequency of
300 kHz.   Lastly, an accelerometer, model 4375, supplied by
Bruel and Kjaer was tested.

The testing required to evaluate the potential replacement
transducers was performed using a Tektronix model 466 storage
oscilloscope connected to the output of the TTA.   An assessment
was made of the relative signal levels of each transducer's response.   The transducers, in turn, were connected to the TTA
amplifier input and mounted on a steel plate.   The size of the
plate was $12 \times 12 \times 1$ in.   Data were recorded for the two types
of stimuli: a light scratch at a diagonal corner of the plate and
light taps at the same location.   A sharp probe tip was used as
a stimulus, with careful attempts to keep the scratches and taps
consistent.   Repeated trials were made in keeping with consistent
data-gathering practice.

The results of the evaluation are shown in Table 6.2.   The
Physical Acoustics model R-15 appeared to have the greatest
sensitivity to a light scratch, by a factor of approximately three
over the other transducers.   This observation from the experimental setup was confirmed when the R-15 transducer was substituted for the original D750 after it failed.   The R-15 was

tested further during a 1000-touch test on a titanium billet. Figures 6.6 and 6.7 show close-up views of the touches made with the previous transducer. No mark was observable during the tests with the new transducer.

Testing of the R-15-I unit was not completed due to the need to modify the TTA amplifier to pick up the signal from the power supply connection to the transducer unit. It is expected, however, that this unit will be successful because of the success of the R-15 unit and the fact that the internal preamplifier should boost the signal-to-noise ratio. This would thus improve the sensor's overall sensitivity.

MACHINING TRIALS ON DISKS AND SPOOLS

Two datums were designed and constructed. These datums were intended to be used in conjunction with specific machining operations on high-pressure compressor (HPC) spools. The datum shown in Figure 6.8 was used for operation 20 on an HPC 3-9 spool. The datum provides reference surfaces for the X and Z axes of the vertical turret lathe. The datum ring was designed to be attached to the part fixture. It was utilized during the proof test of the automated in-process inspection system using the HPC 3-9 spool.

The datum shown in Figure 6.9 has been designed for an application of the system for disk machining. In this application there is an additional requirement for a reference surface at 45 degrees to horizontal along with the surfaces for X and Z.

In anticipation of the machining trials on disks and spools, several improvements were made in the CLM system after the initial experiments had been run. Four changes were made in the system: corrections to the 1050 control software executive, modifications of the TTA, modification of the datum rings, and corrections to the part program. Modifications of the TTA included substitution of the transducer, filtering of the power supply, and a change in the interface between the TTA and the 1050 control.

Changes in the 1050 executive software involved a modification which permitted the part programmer to select various action responses to an out-of-window condition while probing the workpiece. Previously, action was dictated to be an "E stop,"

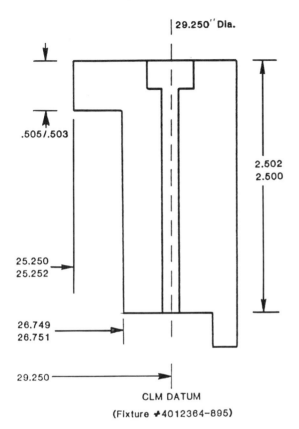

FIGURE 6.8   CLM datum.

which required a complicated operator recovery action. In addition, the executive now includes parametric subroutines and a modification to resolve conversion problems that existed due to the metric scale feedback.

Several modifications were made to the TTA probe that have resulted in an improved signal-to-noise ratio, thus increasing probe sensitivity. Examination of the power supply voltages indicated a substantial amount of electrical noise. Additional filtering was provided to bring this noise level below 40 mV to peak.

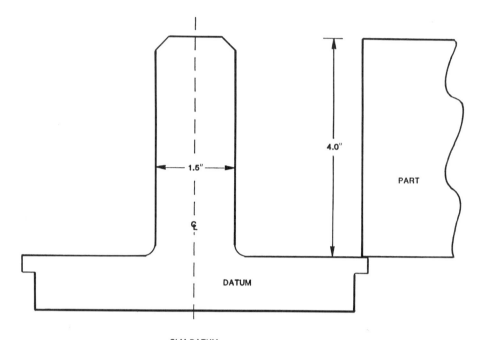

CLM DATUM
(Fixture #4013071-162)

FIGURE 6.9 CLM datum.

As a result of the alternate transducer selection, the more sensitive R-15 transducer was installed in the TTA probe after it was discovered that the D750 transducer had lost sensitivity. After installing the new transducer, a 1000-touch test, as described previously, was performed to verify the operator of the new unit before machining the HPC 3-9 spool.

The interface between the TTA and the 1050 control was modified to improve the reliability of the touch probe operation. The touch signal from the probe was rewired to obtain a reference from -12 volts instead of the previous ground reference. This was in line with the recommendations of the manufacturer of the 1050 control. In addition, the probe "enable" input signal was connected to ground to permit a more positive operation of the "touch" output signal. This allowed the same system operation, since the software polled the touch signal only during a programmed measurement block.

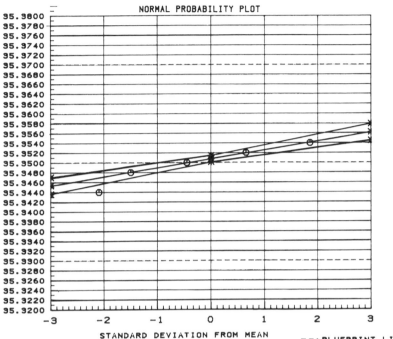

FIGURE 6.10   Normal probability (CLM).

During the machining of the HPC 3-9 spool, a datum reference ring was utilized to measure tool offsets and perform the
final dimensional measurement.   The ring, shown in Figure 6.8,
was mounted on the part fixture and provided a stable reference
for the closed-loop machining measurements.   The part program
for the spool had to be modified to incorporate the use of the
datum ring and to accommodate suggested improvements realized
during previous trials.   During the spool machining operation,
the display time of the deviation information on the 1050 readout
was increased and the part dimensional information was output
from the 1050 through the TTY (teletype) port on the control.

As a result of the successful laboratory experience with the
machining of the HPC 3-9 compressor spool utilizing the CLM, a
production Okuma T lathe was retrofitted with the TTA and linear resolvers in order to obtain CLM performance data in a production environment.   The Okuma lathe CNC control, a Mark

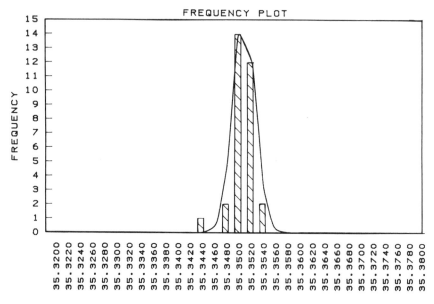

FIGURE 6.11 Frequency distribution (CLM).

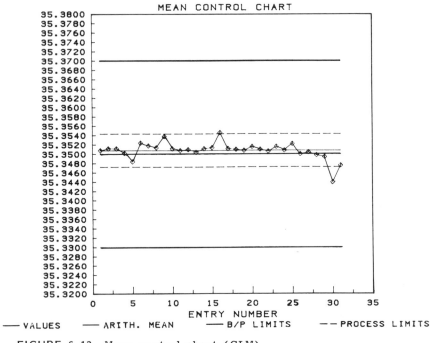

FIGURE 6.12 Mean control chart (CLM).

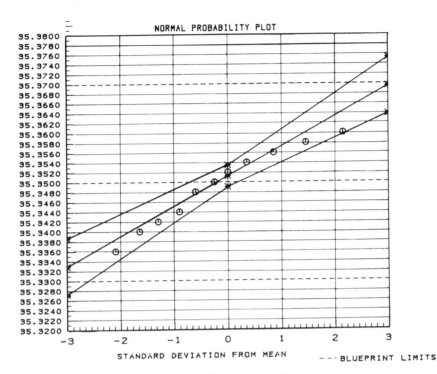

FIGURE 6.13 Normal probability (manual).

Century 1050T, was also modified to accommodate automatic tool offsetting, on-machine part inspection, and reporting of the inspection results. In order to compare the CLM data with existing manual techniques, the same operation was performed on a similar CNC lathe using standard manual inspection tools (verniers, shims, etc.) and techniques to generate tool offsets and final inspection data. The particular machining operation thus compared involved the roughing of a large thin ring part.

The results of the machining study using statistical process control (SPC) tools are summarized in Figures 6.10, 6.11, 6.12, 6.13, 6.14, and 6.15 and in Table 6.3. The dimension measured was an outside diameter of a flange with a nominal value of 35.350 in. The data sets were derived from two separate runs of 31 parts, one set generated using CLM, the other using previously established manual methods. In general, the part dimensions produced by the automatic CLM operation had better than

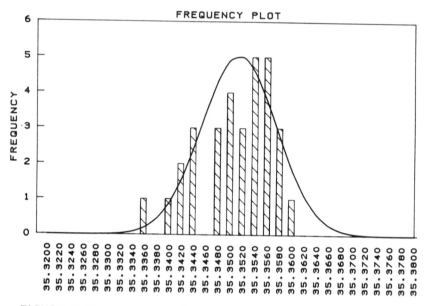

FIGURE 6.14 Frequency distribution (manual).

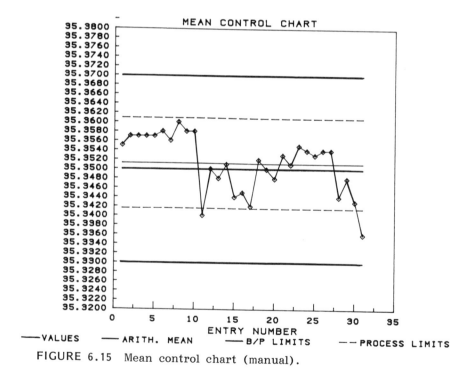

FIGURE 6.15 Mean control chart (manual).

TABLE 6.3   Statistical Results

	CLM data	Manual data
Arithmetic mean	35.3507	35.3512
Mean deviation	0.0011	0.0050
Range	0.0106	0.0240
Variance	0.000003	0.000037
Standard deviation	0.0018	0.0061
Upper control limit	35.3543	35.3609
Lower control limit	35.3472	35.3415

half the spread for process control limits compared with the
manual method.   Table 6.3 lists the statistical results and shows
the dimensional distribution and standard deviations for the CLM
data.   Figure 6.10 is a plot of the normal probability for the
CLM data and Figure 6.11 is a frequency plot of the distribution.   Figure 6.12 shows the mean control chart identifying the
process limits with respect to the blueprint limits.   The corresponding graphs for the data derived by manual methods can be
found in Figures 6.13, 6.14, and 6.15.

As a result of the study, it is expected that the rework
time for this part can be reduced because of the elimination of
incorrect manual offset calculations and incorrect part dimensional measurements.   The scrap rate is expected to be reduced
to zero since the process capability, as shown in Figure 6.10
for the CLM operation, results in all parts produced being well
within the established blueprint limits.   It is anticipated that
similar production parts will experience the same trends in reduction of scrap and rework.

CONCLUSIONS

In conclusion, a true CLM system must have the capability to
automatically measure tool and workpiece dimensions and compensate (offset) accordingly to guarantee that part tolerances are
met.   The system described herein accomplished these basic
functions and included the following additional features:

- Low manufacturing cost.
- East of installation on new and of retrofit on existing lathes.
- No interference with normal machining operations—the sensor is not mounted very close to the cutting action.
- One sensor and associated electronics monitor all tool positions without the need for special tooling.
- Minimal requirements for system training and special part programming considerations.

The acoustic (TTA) approach has numerous advantages and some disadvantages when compared with the touch trigger probe approach. The differences are highlighted in Table 6.4 and are discussed below.

On most lathes, a premium is placed on the fixed number of tool positions available. Adding touch trigger probes to the machine tool for on-machine inspection requires the sacrifice of one or more of these tool positions and an incurred reduction in operating efficiency because more tool changes (reloads of the turret) are required. Using the touch feature of the TTA not only frees up tool positions but also provides improved efficiency because its probe stylus is the tool itself and does not need to be indexed in and out like the touch trigger probe. The single acoustic sensor is less expensive than the two touch trigger probes that would be required in a similar application and is less susceptible to damage from abuse in the shop environment than the stylus of the touch trigger probe.

No design effort is required to provide the probing stylus for the TTA system because the tool used for machining is itself the stylus. The touch trigger stylus, on the other hand, must be carefully designed to access all surfaces and dimensions of the parts to be processed.

Certain liabilities, however, exist with the acoustic approach with its use of the tool as a stylus. If the sensor fails to sense a touch for some reason, the tool would proceed into the part and potentially damage it. To reduce this risk, the TTA system is checked before each critical probe cycle by touching a datum.

It should be noted that if the direct measurement of a thickness is desired rather than a differential measurement with two tool contact positions, a stylus must be used. Here the TTA

TABLE 6.4 Tool Touch Versus Touch Trigger Probes

Tool touch	Touch trigger
(+) No tool positions required	(−) One tool position per probe
(+) Less probe cycle time	(−) Greater probe cycle time from tool (probe) changes
(+) Stylus (tool) to part surface compatibility	(−) Part surface access depends on stylus design
(+) One sensor provides tool and part inspections	(−) Two probes required, one for tool inspection and one for part inspection
(+) Direct calibration from datum, no compliance	(−) Indirect calibration from datum; incurs inaccuracy from stylus-to-stylus compliance
(+) No stylus vulnerability	(−) Delicate probe stylus vulnerable to shop environment
(−) Potential part and tooling damage from overtravel (failure to sense contact)	(+) Small liability from overtravel (crashes)
(−) Thickness measurements may require probing with two different tools or a stylus	(+) Direct measurement of thickness dimensions
(−) Susceptible to spurious acoustic signals	(−) Susceptible to acceleration-related false trips

loses some advantage because a stylus must occupy a position in the turret.

Implementation of the CLM technology has increased machining efficiency, as indicated by shop cost reductions of more than 10%. Furthermore, manufacturing losses were reduced approximately 13% by eliminating measurement computation errors.

These aspects provide a substantial justification for implementing CLM, but perhaps the most significant benefit is intangible. The popular goal in industry today of automating the factory or "lights-out" manufacturing cannot be accomplished until each individual process is automated. Closed-loop machining, as described here, represents a pivotal step toward automating machining, which is so critical to manufacturing today.

REFERENCE

1. D. G. Flom, "Manufacturing Initiative for Advanced Metal Removal," F33615-80-c-5057 (AFWAL-TR-85-4044), May 1985.

7

Sources of Error in Machining Centers and Corrective Techniques

ERIC KLINE and W. ANDREW HAGGERTY / Cincinnati Milacron, Cincinnati, Ohio

INTRODUCTION

Quality products are no accident. Good people get the job done by using both great design and great manufacturing processes.

This chapter presents information about a specific type of manufacturing process, the four-axis horizontal machining center as shown in Figure 7.1. This machine tool type is in widespread use in batch manufacturing factories such as those involved in aerospace, off-road equipment, and precision metal products of all types. Batch manufacturing factories are the most common metal-cutting type.

The four-axis horizontal machining center is highly productive and is capable of automatically producing high-precision results. It is an excellent example of a complex metal-cutting operation that involves complex fixturing, NC programs, and tooling selections. All of these process variables influence the precision results, and there is a great deal to know about the process. Often, this process has to be fine-tuned.

More specifically, the information in this chapter is about troubleshooting the general dimensional accuracy performance of the four-axis horizontal machining center. However, this

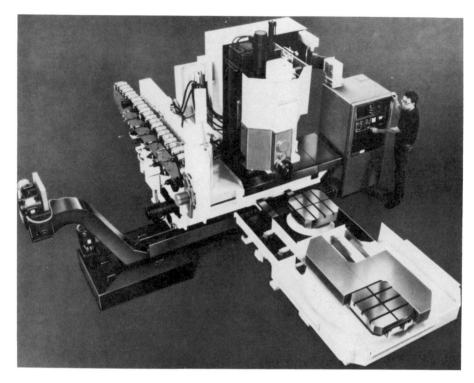

FIGURE 7.1   Four-axis horizontal machining center.

specific accuracy knowledge is generally useful in that it can
be applied to troubleshooting the accuracy performance of other
machine tools as well.

In general, the influences which control the dimensional
accuracy performance of machine tools are a complete set of interrelated sources.   In the abstract, they are an intimidating
group.   Figure 7.2 shows a machine tool designer's checklist of
factors affecting or controlling high-accuracy machining results.
Designers will take into account, to the degree that their cost
targets will allow, all of the machine and control system factors
shown in the uppermost box.   These are the accuracy factors
over which they have control.   Their efforts will influence the
accuracy performance of the machine tool.   Workpiece accuracy,
however, will also depend on the additional factors of environmental effects and operating methods, which are not usually

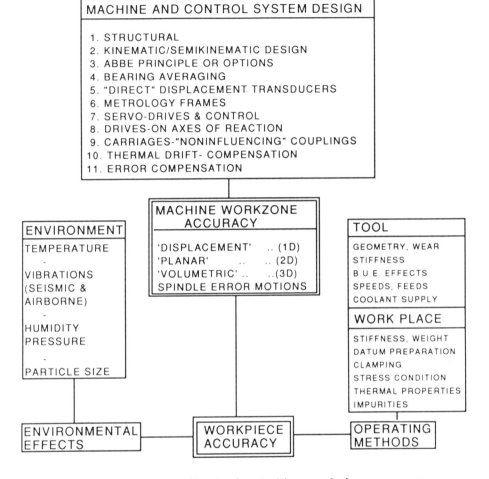

FIGURE 7.2 Factors affecting/controlling workpiece accuracy. (Courtesy of Cranfield Precision Engineering, Ltd., Ref. 3.)

under the designer's control. All of these influences, their interrelations, and which ones are significant in a particular situation do not come easily into the minds of the average production work force.

Recent development work by the author while monitoring
production results from four-axis horizontal machining centers
brings into better focus the influences that are most significant
for this type of machine tool. These influences are controllable
much more accurately and automatically than was previously
thought possible. In most instances, accuracy performance is
better controlled than with the previous standard of production
excellence, the manually operated horizontal boring mill. These
influences will be outlined and discussed in this chapter.

A TYPICAL REPRESENTATIVE PROBLEM

We will look at two examples of average complexity in a workpiece, fixture, and tooling situation for four-axis horizontal
machining centers. In Figure 7.3, a tombstone fixture is shown
carrying two workpiece types. Nearest the machine spindle, a
mechanical valve body is shown during its machining sequence.
Close tolerances are required on the hole and slot sizes and the
spacing between features on this valve body. In Figure 7.4,
an automobile front steering hub is shown surrounded by hydraulic fixture clamps. Close tolerances are required on the
size and location of several bored holes which are at different
angles on this hub.

Suppose there is an accuracy problem with one or both of
these workpieces after machining. Such a problem may create a
production crisis situation. If a manufacturing engineer is as-
signed to troubleshoot this situation, where should the engineer
start?

Typically, one starts by asking for facts and ideas from the
people who have been involved in processing the workpiece up
to the present point in time. It has been the author's experience that if this is done, one gets several different opinions,
as to probable causes and what to try, from each person asked.
Every supervisor and machinist has a well-meant opinion or sug-
gestion, usually a different one.

This approach often adds uncertainty to the problem situation. Several suggestions are made, and it is not clear which
suggestions are likely to have the highest merit. Nor is it very
certain that the suggestions affect the most significant causes
of inaccuracy. One wants to confirm the most significant causes
quickly. A confused situation is easily created because the

FIGURE 7.3 Mechanical valve body on tombstone fixture.

problem is found to have many aspects. There is a need for
a better approach, one where prior experience and knowledge
provide a road map for quickly examining a vital few specific
accuracy factors.

FIGURE 7.4 Automobile front steering hub.

DEVELOPING AN ACCURACY TROUBLE-SHOOTING ROAD MAP

Clearly, a productive solution approach for solving an accuracy problem is a systematic one that begins by examining in logical order the most likely causes. How could a generic list of causes (as shown in Figure 7.2) be sorted and sifted into a more practical road map? The author formulated a good solution from the following ingredients.

Precision engineering is a specialized science in which a great deal is known about the general causes of dimensional inaccuracy. The author has found it helpful when accuracy troubleshooting to think of the general causes of inaccuracy as arranged in the following four groups:

1. Geometric: Precision engineering has very specific knowledge and definitions of geometric sources of inaccuracy in machine tools (Appendix A).
2. Thermal: Precision engineers have a great deal of personal experience with environmental thermal effects on accuracy which is case history specific and less well documented.
3. Process: Precision engineers understand that, beyond geometric and thermal causes, there are important additional process-specific influences on accuracy.
4. Measurement uncertainty: There is uncertainty about accuracy in the very attempt to measure it. Measuring instruments themselves are not perfect.

Add all these influences together for a specific process, such as four-axis horizontal machining, and with careful observation the breadth of influences and a pattern of their relative importance can be constructed.

The teachings of J. M. Juran provide an excellent technique called the "Pareto principle" (Ref. 1) for prioritizing a problem situation that consists of many parts. When it is used with experience and judgment, problem complexity can be sorted and sifted into an organized prioritized list of influences. Precision engineering knowledge and experience combined with Juran's Pareto principle aided the author's observations and analysis of accuracy performance of four-axis horizontal machining centers. Construction of a helpful road map in the shape of a "pyramid of influences" is possible.

THE PIE CHART PYRAMID OF ERROR SOURCES

The road map representing the hierarchy of error sources for four-axis horizontal machining centers is constructed in the form of a set of three charts as illustrated in Figures 7.5, 7.6, and 7.7. This set forms a "pie chart pyramid of importance." Table 7.1 provides a key to the abbreviations used in the charts.

MACHINE VS. PROCESS VS. UNCERTAINTY OF MEASUREMENT

The top of the pie chart pyramid shown in Figure 7.5 is the troubleshooting starting point. All significant factors affecting

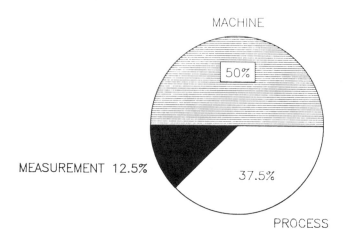

MACHINE

50%

MEASUREMENT 12.5%

37.5%

PROCESS

FIGURE 7.5 Sources of inaccuracies in machined workpieces from four-axis machining centers.

machining center accuracy are mapped into three primary groups:

1. Error sources in the machine
2. Error sources in the process
3. Error in the uncertainty of measurement

Within these three groups, the significant causes for any specific case of inaccuracy will be found. In the interest of thoroughness and efficiency, it is important during initial troubleshooting stages to plan an investigation into each group.

Often at the start of accuracy troubleshooting, the machine is automatically suspected to be at fault. In the author's experience, as shown in Figure 7.5, it is wise to weigh nonmachine causes, i.e., errors in the process and in the uncertainty of measurement, with equal suspicion in an average situation.

It is important to realize that traditional testing of machine tool positioning accuracy and alignment is not a sufficient measure of machine tool accuracy performance. The inaccuracy influences from the process and from the uncertainty of measurement are too great to be overlooked.

It is a surprise to most production personnel that uncertainty of measurement is a significant error source. Measurement

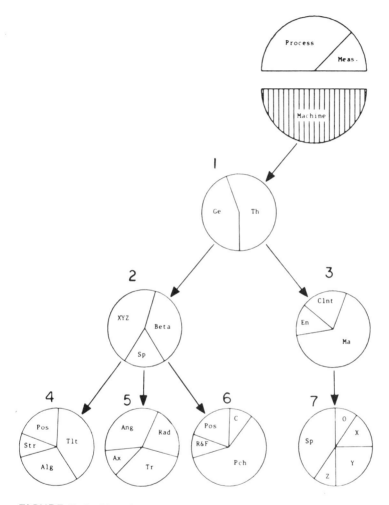

FIGURE 7.6 Machine sources of inaccuracies in four-axis machining center workpieces.

results from dimensional measurement devices such as vernier calipers, micrometers, dial bore gauges, or coordinate measurement machines (CMMs) are often taken as gospel. Nothing could be further from the truth.

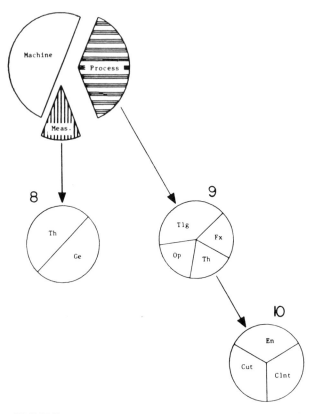

FIGURE 7.7 Measuring and process sources of inaccuracies in four-axis machining center workpiece.

The three primary groups of Figure 7.5 form branches which lead to more specific detail about each group in Figures 7.6 and 7.7.

Error Sources in the Machine (Geometry vs. Thermals)

The road map extension in Figure 7.6 illustrates greater detail about the machine sources of inaccuracies. The second level (Pie #1) of road map detail introduces two principal machine tool error groups: geometry effects and thermal effects.

TABLE 7.1 Key to Pie Chart Errors

Alg	Linear axis alignments (squarenesses)
Ang	Spindle angular runout
Ax	Spindle axial runout
Beta	Rotary axis (table)
C	Beta axis centrality
Clnt	Coolant thermal errors
Cut	Chip cutting thermal errors (machining)
En	All other environment thermal errors
Fx	Choice of fixturing and locating surfaces
Ge	Machine geometry
Ma	Machine thermal errors
O	All other machine thermal errors
Op	Choice and sequence of machining operations
Pch	Beta axis pitch
Pos	Linear and Beta axes positioning errors
Rad	Spindle radial runout
R&F	Beta axis rise & fall
Sp	Spindle geometry and thermal error
Str	Linear axis straightness
Th	Machine thermal errors
Tlg	Choice of tooling influences
Tlt	Linear axis tilts (roll, pitch, yaw)
Tr	Spindle axis tram
XYZ	Linear X, Y, Z axes

In the average troubleshooting situation, static geometric effects (commonly called alignments) of the machine will often be suspected by production personnel as a primary cause. In the author's experience, the dynamic thermal effects are likely to predominate over static geometry effects. This is especially true for machines not equipped with any specific built-in temperature control capability.

Dynamic Thermal Effects

Dynamic thermal effects are error sources that change with time due to changes in operating temperature. They are caused by disturbances from thermal heating or cooling. Complex temperature gradients or differences in temperature between machine elements and between machine, fixture, and workpiece create thermal inaccuracy distortions which dynamically change the geometric alignments.

Thermal disturbances can be sorted into two groups:

1. Heat sources generated within the machine itself, e.g., spindle bearings, servo motors, and hydraulic actuators.
2. Heat sources external (environmental) to the workings of the machine, e.g., coolant, heat from other machinery and sunlight.

These are illustrated as machine (Ma), coolant (Clnt), and all other environmental (En) in level 3 (Pie #3) of the thermal branch of Figure 7.8.

Four-axis horizontal machining centers are heavy-duty milling machines. They have high-power spindles, slides, and hydraulics for operation of a rotary table and for rapidly changing heavy tools. They typically have several internal heat sources which contribute to thermal accuracy disturbances.

Examples of environmental heating or cooling sources are the factory environment itself, which rises and falls in temperature during the course of a workday, and the use of metal-cutting coolant, which usually comes into contact with the surfaces of the machine and workpiece fixture.

Troubleshooting procedures called thermal drift checks measure the magnitude of dynamic thermal error sources. In the author's experience with machining centers, internal heat sources within the machine tool are likely to be a larger contributor to thermal error than the environment unless thermal control features are present.

Static Geometry Effects

Production personnel are usually more familiar with machine tool geometry errors than with thermal errors. Geometry errors are more easily and frequently measured as part of machine tool installation or maintenance. This explains why they come to mind first in a troubleshooting situation.

Geometric error for four-axis horizontal machining centers are divided into two groups, as shown in the level 3 (Pie #2) machine geometry branch of Figure 7.6.

1. Linear axes (XYZ)
2. Rotary axes (rotary table B axis and spindle axis)

In an average case of inaccuracy of a four-axis horizontal machining center, the linear axes group contributes just under half of the machine geometry error. The rotary table B axis and the spindle axis account for the remaining error.

The larger contributors to linear axis geometric errors are the angularity or "tilt" group (slideway roll, pitch, and yaw) and errors of angularity between two axes (alignments or squareness). Definitions of these error motions are given in Appendix A. This information is usually a surprise to most production personnel. The traditional culprit for linear axis error is thought to be simple positioning accuracy. Angularity errors and squareness errors are not investigated. In the author's experience with four-axis horizontal machining centers, if simple linear lost motion (backlash) is satisfactory, the remaining inaccuracies of linear position (such as feedback position transducer error) are quite small compared with linear position errors resulting from errors in angularity and squareness.

For typical machining center workpieces, spindle errors are not significant sources of error when compared to geometry errors from the linear axes and from the rotary table (Pie #2). Spindle axis tram and spindle angular runout errors are the larger spindle errors (Pie #5).

The predominant rotary table B axis geometric error is a pitch or tilt error between angular positions of the rotary axis. This is due to the effect of a sizable Abbé offset length (see Appendix A) which cants the fixture and workpiece out of position.

Error Sources in the Process

The road map extension in Figure 7.7 illustrates inaccuracies from the process group in greater detail. These inaccuracies

are the most variable error sources because of the uniqueness
of each fixturing and metal-cutting tooling situation for a given
workpiece (see workpiece/fixturing example in Figures 7.3 and
7.4. These error sources must not be overlooked in a troubleshooting situation. It is wise to treat them with as much suspicion as other error sources in an average situation.

Process errors of four-axis horizontal machining centers
are divided into four groups as shown in the level 2 (Pie #9)
process branch of Figure 7.7:

1. Choice of tooling (Tlg)
2. Choice of fixturing and locating surfaces (Fx)
3. Thermal effects (coolant, hot chips, etc.) (Th)
4. Choice and sequence of machining operations (Op)

The Influence of Tooling

In the process error group, inaccuracies associated with tooling
are the largest source of error. This is no surprise, because
being able to make good choices of metal-cutting tooling is a
skill based on experience.

A case history will give an example of the influence of tooling. Figure 7.8 shows a schematic of a machining center workpiece where accuracy was influenced by a choice of tooling.
The workpiece is a cast-iron compressor housing with a flat
face on which it was necessary to drill two pairs of 0.090-in.-
diameter holes and two 0.250-in.-diameter holes. The problem
was with the machining accuracy of the two pairs of small-diameter holes. Their dimensional location varied too much even
though the mean value was on target. The variation in location
of the two larger-diameter holes was much better and was within specifications.

Production personnel suspected that the machine was positioning erratically. Although this could have been possible,
their thinking proved faulty. It did not explain why there was
a difference in positioning repeatability between the larger-diameter holes and the smaller-diameter holes. The satisfactory
results for the two larger holes correctly indicated that the
positioning performance of the machine was acceptable.

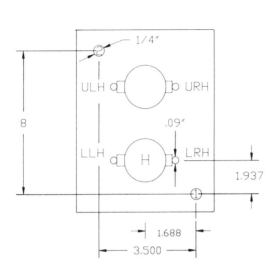

FIGURE 7.8   Diagram of compressor housing.

A recommendation was made to examine the tooling. It was found that standard accuracy and not precision accuracy drills and collets were being used. This was not consistent with the desire to have precision hole location. A tooling change was made. Precision carbide drills and precision collets replaced the standard accuracy tooling.

Figure 7.9 shows the improvement in results obtained from this tooling change. The location repeatability error seen for the small-diameter drills improved by a significant 50%. The better choice of tooling reduced the causes of inaccuracy common to small flexible drills and improved the process to within accuracy specifications. These inaccuracies were outside the control of the machine positioning. The influence of the new tooling produced a smaller improvement (20%), as expected, on the more rigid larger-diameter drills.

The Influence of Fixturing and Location Surfaces

It is often a challenge to provide a fixture for a complex workpiece. The fixture must be noninterfering with machining operations. It must be nondistorting to the workpiece while at the

| Drill Diameter | HOLE LOCATION REPEATABILITY (6 Sigma) Tooling Choice | | IMPROVEMENT |
	Std	Spec	
.250"	.0025"	.0020"	20%
.090"	.0050"	.0025"	50%
	Std HSS Drill Std Collet	Solid Carbide Precision Ground Drill Precision Collet	

FIGURE 7.9   Tooling influence on workpiece accuracy.

same time clamping it solidly enough for the workpiece to with-
stand the high forces of machining.   Fixture and locating surface
selection, like tooling selection, is a skill based on experience.

From an accuracy standpoint, important decisions about fixturing are linked to provisions made in the design of the workpiece for manufacturability.   A good workpiece design will include provisions for proper locating surfaces for processing.
But even with good locating surfaces, the processing approach
can still go astray.   The following case history is an interesting
example involving the misuse of fixture locating surfaces.

A schematic of a machining center workpiece where accuracy
was influenced by a poor choice of fixture locating surface is
shown in Figure 7.10.   The workpiece is the cast-iron compressor housing shown before in Figure 7.8.   This time, however,
the workpiece feature of interest is an angled slot coincident
with the centerline of a previously bored hole.   The problem
was that the dimensional location of the slot varied too much
with respect to the target location, the horizontal centerline of
the bore.   Production personnel suspected that the machine was
positioning erratically or that there was some problem with the

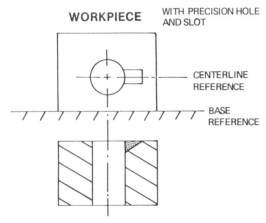

FIGURE 7.10   Diagram of compressor housing.

tooling. These could have been possible, but the real cause
proved to be related to the fixture and locating surface.

The steps associated with machining of the slot were examined. It was learned that the location of the slot was being
established by reference to a locating surface on the base of
the workpiece and not to the centerline of the bored hole itself.
This approach allowed any variation between the hole centerline
and the base surface to become part of the location error between the slot and the hole centerline. This indirect choice of
locating surface was very inappropriate for this precision machining operation.

A change was made in the processing steps for machining
the slot. The hole centerline became the new locating surface
for machining the slot. Figure 7.11 displays the improvement
in results from this change. The mean deviation in slot position with respect to the bored hole improved from +0.0020 in.
to -0.0001 in. and there was a 6-to-1 narrowing of the
repeatability.

Thermal Effects (Coolant)

The use of coolant for metal-cutting applications is very common
and necessary. The cutting edges of the cutting tools require
coolant to keep from burning up and to lubricate the interface
between the tool and the workpiece. The flow of coolant flushes
produced chips away from the cutting zone. Without coolant,
very hot chips and workpieces would cause handling problems.

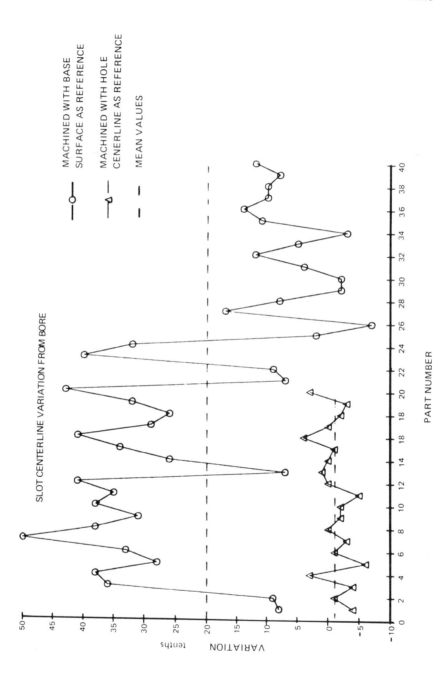

FIGURE 7.11 Chart of process capability.

Workpiece temperature is highly influenced by the temperature of the coolant. Therefore, the exact size of the workpiece
is highly influenced by the use of coolant. This is especially
true for aluminum workpieces, since this metal has a high coefficient of thermal expansion. For the highest-accuracy cutting
operations, it is necessary to control the temperature of the
workpiece. This can be accomplished by regulating the temperature and the flow of the coolant.

Based on the machining center experience of the author, it
is recommended that the temperature of the coolant be regulated
to 1 or 2°F below ambient air temperature and that the flow of
coolant wash over the entire workpiece and fixture.

Choice and Sequence of Machining Operations

The NC programming of four-axis horizontal machining centers
is a very complex task. Early in this process, the machining
operations and their sequence are selected. Most of the subsequent effort goes into defining the geometry of the NC tool path.
However, assuming that the tool path is correct, the precision
results will be most influenced by the choice and sequence of
machining operations.

The goals in choosing the type and sequence of machining
operations for high-accuracy results are (1) to keep machining
forces and temperatures low, (2) to give machining forces a
favorable orientation, and (3) to keep machine positioning moves
headed in the same direction. Such choices are case specific
but a few practices are known to be helpful:

1. All rough machining operations should be completed prior to
    finishing operations. This allows any internal stresses with-
    in the workpiece material to be rearranged into a final pat-
    tern before final cuts are made.
2. A pause between rough and finish operations is helpful so
    that the warm temperature of the workpiece, fixture, and
    machine can decrease.
3. Finishing operations should remove minimum amounts of
    stock to keep cutting forces and tool deflections small.
4. Machine positioning motions for all related finishing opera-
    tions, such as bored holes with close center distance toler-
    ances, should approach the workpiece from the same direc-
    tion, e.g., from the left and up, and this should be done
    in succession and with a minimum of elapsed time.

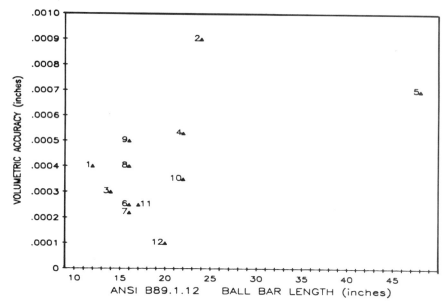

FIGURE 7.12   Advertised CMM volumetric accuracy.   1:   Brown
& Sharpe Validator PCR 500 Series (0.0004 in./12-in. Bbar);
2:   DEA Bravo 2104/1 Meas Robot (0.0009 in./24 in. Bbar);
3:   DEA Bridge Type (0.0003 in./14 in. Bbar);   4:   DEA Bridge
Type (0.00053 in./22 in. Bbar);   5:   DEA Bridge Type (0.0007
in./48 in. Bbar);   6:   Zeiss Mauser Bridge Type (0.00025 in./
16 in. Bbar);   7:   Zeiss PMC Bridge Type (0.00022 in./16 in.
Bbar);   8:   Federal Meas Robot (0.0004 in./16 in. Bbar);   9:
Sheffield Appolo Series (0.0005 in./16 in. Bbar);   10:   Johanson
Cordimet 801 (0.00035 in./22 in. Bbar);   11:   Mitutoyo FN 905
(0.00025 in./16 in. MCG);   12:   Moore M-60 (0.0001 in./20 in.
Bbar).   1986 IMTS, 1987 Q'EXPO

Error in the Uncertainty of Measurement

The second part of Figure 7.7 (Pie #8) addresses an issue fundamental to all dimensional measurements: how accurate are the
devices doing the measuring?   These devices, such as vernier
calipers, micrometers, dial bore gauges, and CMMs, are expected
to be of high accuracy, but in practice they are far from being
perfect.   In fact, the tighter the workpiece tolerances being measured, the more likely it is that the inaccuracies of measurement
will become a significant fraction of the measurement being made.

In 1985, the inaccuracies of measurement related to CMMs were better recognized when the American Society of Mechanical Engineers published a report concerning methods of evaluating coordinate measuring machines (Ref. 2). This standard includes a revolutionary section dealing with "volumetric" accuracy performance, which had not been previously addressed in American standards. Volumetric accuracy is a much better overall measure of CMM inaccuracy than previous single-dimensional methods. This has a significant value to users of these machines. With this standard, it is now easier for CMM users to recognize and compare CMM accuracy capabilities and limitations. Some advertised volumetric performance data are shown in Figure 7.12. The data were collected by the author at two trade shows during 1986 and 1987.

When workpieces are measured, it is very common to interpret inaccuracies observed as belonging entirely to the workpiece. This can be true only if the measurement devices are perfect, which is almost never the case. The inaccuracy of a measurement device is always some fraction, hopefully a small one, of the magnitude of the measurement being made.

Ideal metrology and precision engineering practice calls for the accuracy of a measuring instrument to be 10 times better than the tolerance band being measured. For example, if ±0.005 in. is the desired tolerance of a workpiece dimension, the measurement device should be certified to have an accuracy of ±0.0005 in. or better. Frequently, the pressures of meeting production schedules or of keeping capital expenditures for measurement equipment low lead to relaxation of the 10:1 rule and settlement for ratios of 5:1, 3:1, or worse. This greatly increases the uncertainty in the measurement accuracy, produces corrupted measurement data, and passes or fails workpieces incorrectly.

There is a simple two-step test that indicates the magnitude of the measurement uncertainty present in a given situation.

1. Take repeat measurements of the same workpieces on the same measurement device. This will measure the repeatability of the measurement device and of the measuring setup and measuring technique. Ideally, the repeatability observed will be less than one-tenth of the workpiece tolerance being measured.

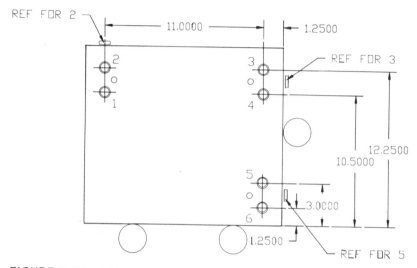

FIGURE 7.13   Diagram of CMM test part.

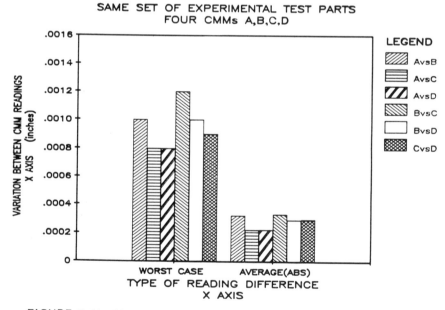

FIGURE 7.14   Measurement variation between several CMMs.

2. Take repeat measurements of the same workpieces on several different measurement devices. This will measure the disagreement between the measurement devices. Ideally, the disagreement observed will be less than one-tenth of the workpiece tolerance being measured.

The author used this measurement of uncertainty approach to learn what values are representative for a group of four CMMs located at four different plant sites. Three of the four machines were older manually operated CMMs and one was a newer CNC CMM. Eleven experimental test parts, as diagrammed in Figure 7.13, were measured three times each with identical setups on each CMM. Figure 7.14 shows results for the X-axis direction. The worst-case X-axis disagreement between all the CMMs was 0.0012 in. and was for a specific bored hole location. The average disagreement was about 0.0003 in. Based on the results for these CMMs, it would be questionable to use these measurement devices for measuring workpiece tolerances tighter than about 0.003 in.

In certain circumstances, CMM accuracy performance can be enhanced in a very practical way. This is accomplished by using the CMM as a comparator, i.e., known length standards, such as length rods or master workpiece models, are measured (in similar spatial orientations as the workpiece) by the CMM along with the workpiece features. Then, the differences in measurement are analyzed between the length standards/models and the workpiece features, instead of analysis of the absolute CMM measurements.

CORRECTIVE TECHNIQUES

The large number of error sources collectively contributing to the overall accuracy of a workpiece may seem overwhelming. In a particular case, however, it is likely that only two or three will cause 80% of the problem.

It is also helpful when troubleshooting accuracy problems to be aware of earlier experiences of precision engineers that have shown that machine errors are deterministic; i.e., they obey cause-and-effect relationships and their behavior is repeatable for repeatable circumstances. Error sources are not mysterious random events or probabilistic events uncontrolled by the laws of science. The laws of science do apply, and assignable causes for inaccuracies are present in any given situation.

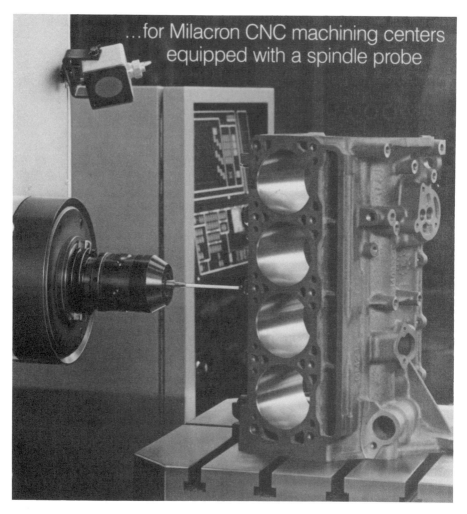

FIGURE 7.15   Touch trigger spindle probe.

Knowing the laws of science and being able to track down causes of inaccuracy does not, unfortunately, always lead to practical efforts or economical fixes for these causes. Sometimes finding a way to compensate for the effects of error is easier than correcting or fixing the error itself.

CLOSING THE PROCESS LOOP

The four-axis horizontal machining center is a manufacturing process. Workpiece materials go into the machine, cuts are taken in a sequential manner, and machined workpiece materials come out. Accuracy checks are sometimes made by an operator interrupting the process but most often are accomplished afterward; a traditionally open-loop process would benefit from an automatic internal gauging means to better control significant portions of the accuracy variation in the process. The success of this idea has already been proved on grinding machines, which frequently operate with a closed-loop capability using in-process gauging.

A new feature available on most machining centers is the precision spindle probe. It enables the machining center to have in-cycle comparative gauging capability similar to the in-process gauging for grinding. The distinction between in-cycle and in-process gauging is that in-cycle means intermittent action taken between sequential cuts while in-process means continuous action taken during grinding.

Figure 7.15 shows a typical machining center precision spindle probe. It has a touch trigger; i.e., when the workpiece is touched with the end of the probe stylus, the probe triggers a reading of the machines' current multiaxis position. In the author's experience, the repeatability of the probe system as applied to machining centers is about ±0.0002 in. with a 50 mm length stylus. This capability enables the machine to determine with reasonably high precision the comparative axis location of workpiece geometric features.

THE BENEFITS OF IN-CYCLE PROBING

The accuracy of most features of machining center workpieces is defined in relation to the dimensional position of other features of the same workpiece. For example, the positions of holes are specified with respect to one another, or the positions of edges, slots, and surfaces are specified with respect to one another or to holes. Without in-cycle probing, the machine blindly cuts these features into the workpiece following the direction of numerical program data based on the assumption that the machine's positioning performance, the fixturing and tooling performance, and the thermal behavior will all be noninfluencing.

These influences always have an effect, for all the reasons
previously presented. The benefits of in-cycle probing extend
the function of the machine to include a precision sense of touch
and an automatic way to make midcourse error compensations
during the sequence of machining operations. For many sources
of inaccuracy, but not all, these compensations minimize detrimen-
tal inaccuracy effects.

A good way to begin machining operations using the precision spindle probe is to take location measurements of identifying workpiece surfaces prior to any cuts. This enables the
following:

1. Better centering of subsequent machining operations on the
    workpiece, thereby equalizing stock removal and equalizing
    finished wall thicknesses, etc.
2. Tests for a dimensionally proper workpiece: presence, size,
    orientation, etc.

The most common use of the precision spindle probe to improve accuracy is to shift the machining location of workpiece
features to compensate for dimensional shifts in workpiece location caused by all the various process error sources. This
works because most, but not all, workpiece features are dimensionally located with respect to other workpiece features. So,
if the axis location of one workpiece feature shifts or drifts for
some reason, the axis locations of other features, that are dimensionally referenced to it should be likewise compensated.

The compensation is accomplished by utilizing the probe's
precision axis location referencing ability just prior to the machining of each workpiece feature. The axis locations of previously machined features are precisely compared with their programmed positions. Dimensional shifts, if any, are then calculated by simple subtraction.

Once the shifted amount is determined, the machine control
uses the amount as compensation for an identical shift in the
numerical program data for the subsequent precision machining
operation. If done correctly, this will produce machined features that are more accurately located relative to one another
than they would be if no compensation had been applied. A
specific example of the effectiveness of this technique was described earlier in the discussion referring to Figures 7.10 and
7.11. That discussion concerned a change in the processing

FIGURE 7.16   Cast-iron link.

steps for a precision slot where a change in the choice of locating surface, made possible by the precision spindle probe, resulted in a large improvement in accuracy.

The most advanced accuracy technique of the precision spindle probe known to the author at this time extends the accuracy
capability of machining centers to match or exceed that of the
traditional standard of excellence, manually operated horizontal
boring mills. The technique requires additional hardware in the
form of mechanical reference length gauges that are fastened
onto workpiece fixtures. The mechanical reference length gauges
provide in-cycle measurement standards in the immediate vicinity
of the workpiece. A reference gauge length is required for each
precision dimension desired. The reference gauge lengths are
used as master comparators and are probed prior to machining
of the precision dimensions desired.

Figure 7.16 shows an example of this fixture reference length
gauge technique. The workpiece is a cast-iron link part which
has a precision hole located at each end. High accuracy is desired for the center distance separation between the two end
holes. A reference length gauge is mounted in close proximity
to the link on a machining center tombstone fixture. The length
of the gauge and its location on the fixture are precisely known.
The two holes are machined in the workpiece in relation to the
two ends of the reference length gauge. Each end of the length
gauge is probed in-cycle prior to the finish boring operations on
the holes.

	Hole center distance variation range (in.)
Machining center without ref. surfaces	0.0028
Machining center with ref. surfaces	0.0011
Pre-FMS manual boring mill	0.0020

Note: Nominal hole center distances of link workpiece family
ranged from 3.0000 in. to 18.0000 in.

FIGURE 7.17 Flexible manufacturing system machining center
accuracy improvement through the use of precision spindle
probe and multiple fixture reference surfaces.

The accuracy for link parts with this technique on machining centers was about 2-1/2 times better than the accuracy without the technique. See Figure 7.17. More important, the results matched or exceeded those previously obtained with manually operated boring mills. An extra bonus was higher productivity and more consistent results from the machining centers.

APPENDIX A*: A REVIEW OF ACCURACY DEFINITIONS

Orthogonality and Degrees of Freedom

Figures 7-A1 and 7-A2 show orthogonality and degrees of freedom, respectively.

Precision, Accuracy, and Resolution

Precision is often called repeatability. For our purposes, it is defined as an average difference among a group of measurements at a target position.

Assume that a marksman is shooting at a target. The tightness of his shot pattern is a measure of precision; it says something about his repeatability, using that gun, in that environment.

Accuracy is how closely the measurements agree to a given standard for the measurement. Precision, by comparison, is how close the measurements agree to each other.

In our example of the marksman, the closeness of his shots to the bull's eye is the measure of his accuracy. It is possible for all his shots to have been close together (precise), but to have been far from the bull's eye. Precision and accuracy are not the same. See Figure 7-A3.

Accuracy in linear positioning is the difference between the average position of a group of measurements and the location of the target position for those measurements. Accurate parts are exactly like the drawing for those parts. Precision parts are exactly like each other. A batch of parts, all rejected for the same flaw, are precise but not accurate.

Resolution is the smallest increment of scale. In the target example, resolution is the distance between the rings in the target.

---

*Reprinted with permission of Hewlett Packard.

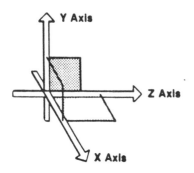

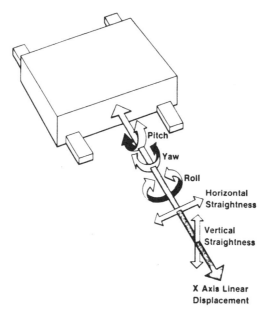

FIGURE 7-A1 (Top)   Squareness between axes (orthogonality).
FIGURE 7-A2 (Bottom)   Degrees of freedom.

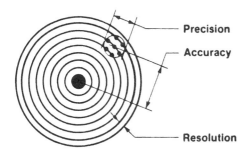

FIGURE 7-A3 Target example of precision, accuracy, and freedom.

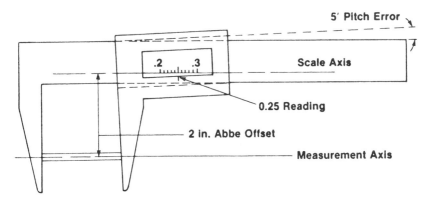

FIGURE 7-A4 Caliper example of Abbe' error.

Abbe' Offset Error

Abbe' offset error can appear in a laser measurement as well as in a measurement made using a more traditional method. It results from the combination of

1. An angular offset in the measurement system
2. A linear offset between the measurement axis and the scale axis in the measurement system.

Abbe' offset error will make the indicated length either shorter or longer than the actual length, depending on the direction of the angular offset, as shown in Figure 7-A4.

The amount of measurement error resulting from Abbe' offset is

Offset distance   tangent (offset angle)

For an angle of 20 degrees or less, the tangent of the angle divided by the size of the angle in arcseconds is between 5.05×10^{-6} and 4.84×10^{-6}. This means that you can get a very close approximation of the Abbe' offset error by multiplying the measurement offset distance by

$$\frac{5 \times 10^{-6}}{\text{arcsec}} \times \text{the angle's size (in arcsec)}$$

The units of the error will be the same as the units of the measurement (and of the offset) (centimetres, inches, etc.), since the tangent value is dimensionless.

Abbe' offset error increases in proportion to the size of the angular offset and the distance of the linear offset, and is really a question of "what are you trying to measure?" For example, if you want to determine the accuracy of a machine tool's leadscrew or scale, you would make your measurement as closely as possible to the screw or scale, which would minimize the results of Abbe' offset. If, on the other hand, you wanted to know how accurately the machine positions its cutting tool, you would make your measurement along the tool's path; this would give you a true picture of the positioning accuracy, and would include compensation for geometric inaccuracies of the measurement system (including those due to Abbe' error).

As an example of Abbe' offset error, consider the caliper shown in Figure 7-A4.

REFERENCES

1. J. M. Juran, "Managerial Breakthrough—A New Concept of the Manager's Job," McGraw-Hill, New York, 1964.
2. American Society of Mechanical Engineers, "Methods for Performance Evaluation of Coordinate Measuring Machines," ANSI/ASME B89.1.12M, 1985.
3. P. A. McKeown, "High Precision Manufacturing and the British Economy," Proc. Instn. Mech. Engrs., 200 76, James Clayton Lecture presented in London 23 April 1986.

8
Machine Vision Applications in Machining and Forming Processes

WALTER J. PASTORIUS / Diffracto, Ltd., Windsor, Ontario,
Canada

Machine vision systems have been installed for several hundred
in-line applications for the inspection of parts produced by metal
machining and forming operations. Many installations have utilized direct feedback to control the process.

This chapter reviews the experiences to date and the underlying technologies represented. Proven sensor solutions are described in the areas of dimension, surface defects, microfinish,
and contour of a wide variety of machined and formed parts. In
each case, a discussion of the past achievements in machine
vision in these areas is presented, primarily inspection of the
finished product. A discussion of future trends, enabling technology requirements, and suggesting R&D efforts concludes the
discussion.

APPLICATION AREAS

The application of machine vision to the efficient operation of
machines breaks down into three areas; gauging and inspection
of the parts produced, checking of the tools themselves, and
use of sensors incorporated in the tools or the machines to sense
part condition or some variable in the actual manufacturing process.

INSPECTION APPLICATIONS

Machine vision has found most applications to date generally at final inspection after the part is relatively clean (and usually washed). However, some checks have been made directly post-process and, recently, in-process.

Similarly, the inspection of tools can be performed postprocess, in-process, or during setup. The location of the tool plus its actual condition is often of interest. Such location is generally determined in-process or preprocess and may be determined from sensors built into the tool itself.

In the inspection of tools or parts, one is basically looking at the size, and in the case of tools this could be a reduced size or altered form due to wear. Also of interest is general condition, that is, present/not present, broken/unbroken, and surface finish or condition (porosity, tool marks, and the like).

A list of the advantages of machine vision for inspection applications is given in Appendix A.

Justifications for Machine Vision

Justification requirements obviously depend on the application. However, in general, machine vision for inspection of parts offers several real advantages: higher speed, freedom from damage, decreased maintenance, and improved accuracy, especially over time.

Productivity

Productivity is generally enhanced because machine vision allows faster inspection with less downtime. This is primarily due to the freedom from breakage and the ability of the sensor to operate at high accuracy without dirt buildup or wear on the sensing device. The lack of any requirement for gauge probe actuation (which is required with contact-type outer diameter gauges, for example) is also a maintenance feature.

From the point of view of capital costs, the vision gauge has historically been more expensive. However, as volumes rise, the per-sensor cost of inspection is dropping. Indeed, several factors can reduce the effective cost of vision below that of a conventional gauge:

1. Ability to handle a higher volume of parts, so that it may
    be sufficient to buy one gauge instead of two.
2. Ability to check many more features of a part, so that
    fewer gauges may be needed.
3. The increasingly demonstrable ability to run without mas-
    ters over large ranges, which decreases the cost and time
    of mastering as well as the stocking of the master compon-
    ents.

Quality

In the area of quality, when comparing a contacting to a noncontacting gauge, the same dirt buildup can be a factor if contacts are not cleaned. Moreover, the vision gauge almost invar-
iably gives more and better data, generally in a flexible manner
requiring fewer changeovers. It is also typically based on accurate digital sensor arrays which are linear and driftless.

For example, the all-time record on a complete final inspection vision based automatic crankshaft gauge holding a master
reading is 14 days on all dimensions to 40 microinches (1 micrometer) with no rezeroing. This is virtually impossible with conventional gauges. Indeed, one can now argue that gauging
capability is catching up with the accuracy capabilities of the
latest machine tools.

Range

Vision-based checks typically have more range for any given
accuracy level than checks with conventional linear variable
differential transformer (LVDT) probes on air gauges. This
leads to greatly enhanced flexibility.

Visual Checks

For measurement of other part variables such as defects and
surface finish, the machine vision solution has no equal in conventional practice other than manual methods, with which there
is virtually no contest. The vision device is almost a certain
payback compared with human labor, assuming it can do the job
reliably. This means making a correct call on virtually all the
bad parts while rejecting only a few percent or less of the good
parts as "false rejects."

The quality improvement is dramatic, since no human can, in normal factory conditions, catch more than 85% of the defective parts. Typically, humans are 60 to 85% efficient, whereas the vision gauge, properly constructed, is virtually 100% efficient. In addition to such performance, the vision gauge can be justified in terms of reduced labor cost, scrap, and warranty expense plus increased customer satisfaction.

THE TECHNOLOGICAL STATE OF THE ART

It is now of interest to summarize the state of the technology as known by our company—that is, what has been achieved in practice from the sensor point of view.

The machine vision technologies utilized are triangulation ranging, back-lit imaging of outer diameters or edges, and reflective imaging (front illumination) generally for part surface or defect condition. The principles are illustrated briefly in Appendix B.

Measurement of Outside Diameters

It has been proved in numerous applications that OD measurements to accuracies of 10 to 20 microinches (0.25 to 0.5 μm) can be made reliably in-line on the normal run of machined metal parts. This includes the accounting for dirt, wash marks, coolant, etc. that still remain on the parts after wash, blow off, etc.

Figure 8.1 illustrates inspection of an automotive four-cylinder crankshaft for all final dimensions, journal diameters, out of round, oil seal and pin diameters, etc. In addition, vision-based sensors determine keyway location, thread depth in the flywheel bolt holes, and thrust wall runout and width.

Forty-nine camera units are employed. They operate through five real-time microcomputers that process the data with a programmable logic controller (PLC) providing overall machine control. Extensive efforts have gone into this gauge, making it able to read reliably to 12 millionths of an inch (0.3 μm) in the presence of normal ambient factory dirt and washer residue.

The sensors stand off 6 in. from the part surface and allow the part to be transferred without requiring the sensors to be actuated, increasing speed and improving accuracy. This ability

FIGURE 8.1 Automatic vision-based crankshaft gauge.

FIGURE 8.2 Vision-based programmable turbine blade inspection system (PACS).

259

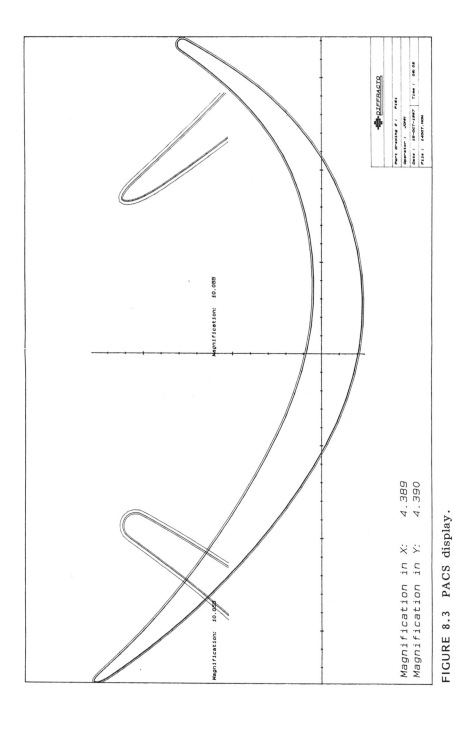

Magnification in X: 4.389
Magnification in Y: 4.390

FIGURE 8.3 PACS display.

FIGURE 8.4   Vision-type gear gauge for rear axle hypoid pinions.

to read accurately from a large standoff is a basic advantage
and makes it much easier to locate sensors near machines.

Measurement of Contour and Thickness

Triangulation measurements have been used extensively to contour parts.   Figure 8.2 illustrates a turbine blade inspection
machine which is capable of contouring a complete range of turbine airfoils.   A graphic display is shown in Figure 8.3.   These
machines are all utilized in a direct CAD interface mode, increasingly in a closed loop to the machines.   The turbine blade inspection machine is substantially faster than other coordinate-
measuring approaches, and the small stylus "point" of the laser
spot can contour the critical leading and trailing edges of the
blades much more easily.

FIGURE 8.5   Model 300 noncontact CMM probe.

Figure 8.4 illustrates a further application of high-accuracy scanning technology, in this case to a high-speed flexible production gauge for a complete family of rear-axis hypoid pinion gears for tooth space error and pitch line runout.

Not all machines need be so large or special purpose. Figure 8.5 shows a triangulation-based sensor probe provided for general application to coordinate measuring machines and the like. It has an accuracy of 0.0002 in. (5 μm), a range of 0.4 in. (10 mm), and takes 200 readings per second (typical).

FIGURE 8.6   Valve stem defect inspection.

Surface Defects

Another major application of machine vision to machined parts
is in inspection for surface defects, a direct replacement of
manual visual inspection.   The surface defects in this case are
primarily those which show up after the machining process,
such as porosity and tool marks.

Figure 8.6 illustrates an interesting application of vision
technology to cylindrical part surfaces, in this case to an automotive engine valve stem.   The complete stem is inspected in
300 milliseconds by imaging a line of the stem surface onto a
linear photodetector array which makes thousands of scans per
second.   Rotating the part allows the total surface to be mapped.

Figure 8.7 illustrates a similar application to a disk brake
rotor.   Here the scan is radial, with the part or sensor rotated.
At lower resolution, this can be done with fixed two-axis scanning arrays or television cameras as well.

FIGURE 8.7 Disk brake rotor defect inspection.

A bore defect application is illustrated in Figure 8.8, which shows a close-up of the sensor unit used for sensing defects and features in the piston bore of rack-and-pinion power steering housings. Similar probes have been built for rifle barrels, pinion gears, master brake cylinders, engine cylinder bores, etc.

FIGURE 8.8 Power steering housing bore inspection probe.

Microfinish

A field related to discrete surface defect detection is the determination of surface microfinish. In this case, the optical sensor
generally used is based on the distribution of light scattered by
the surface. This technique has proved extremely successful in
in-line applications where all parts inspected are produced by
the same process, which is accordingly monitored with the gauge.
The technique works best on ground, honed, lapped, cold-rolled,
and other surfaces with random spacing between peaks and valleys in the range 0 to 64 AA. Numerous applications, however,
exist on roller burnished surfaces and turned surfaces up to
85 AA.

APPLICATIONS TO FORMED SHEET METAL
AND PLASTIC

Generally speaking, applications to formed metal or plastic parts
have paralleled those in the metal machining area, namely inspection of the incoming material and inspection of the resultant
parts for dimension and defects. Parts inspection falls into four
basic categories:

1. Dimensional checks of panels in-line after press working or
   molding operations.
2. Programmable dimensional checks off-line.
3. Surface condition checks of formed surfaces or incoming
   stock.
4. Checks for hole presence and attached parts if any (weld
   nuts).

Other applications include measurement of dies, molds, die
models, tooling aids, and other tools used in these processes.

Specific Sheet Metal Gauging and
Inspection Applications

In-Line Gauging of Body-in-White
and Components

Another important sheet metal gauging application is the 100%
dimensional gauging of the body-in-white and the components
thereof, especially side frames, doors, and underbody. Historically, such checks have been accomplished with electronic touch
probes, but this is giving way to noncontact vision sensors
which offer continuing accuracy, noninterference with the process, and freedom from crashes, distortion, and associated
downtime. In addition, certain features such as hole and slot
locations can be much more easily determined. Where required,
stud locations and other special features can also be reliably
detected. Systems provided are generally complete with user-
friendly graphics, statistical packages, and intergauge communication.

  Figure 8.9 illustrates a large gauge performing 130 checks
on a body-in-white. Data from all points are taken to 0.004 in.
(0.1 mm) in less than 1 second.

FIGURE 8.9   Vision-based gauge for body-in-white.

Surface Inspection

Our company has been working since 1977 on the problem of
sheet metal surface inspection.   In the course of this work, an
important discovery was made, now called "DiffractoSight."   In
this case a picture is worth a thousand words.

Figures 8.10 and 8.11 are photographs of car body sides
taken with and without this optical technique, which greatly
magnifies local geometric contour deviations such as dings, pim-
ples, low spots, buckles, waves, and sinks.   The technique

FIGURE 8.10   Diffracto sight image of body side (with).

has been successfully tested on dies, molds, and wood models, as well as both painted and unpainted (highlighted) metal and plastic.

Incorporation of Sensors with Machine
Tools for Control Thereof

It is of interest to consider whether machine vision sensors can be used in a more direct manner to control the process. Some aspects of this are now discussed.

INSPECTION OF CUTTING TOOL CONDITION

Inspection of tools basically includes location and/or dimension and defects such as chipped or broken edges. Linear parameters include length, diameter, or radius; alignment; and cutting

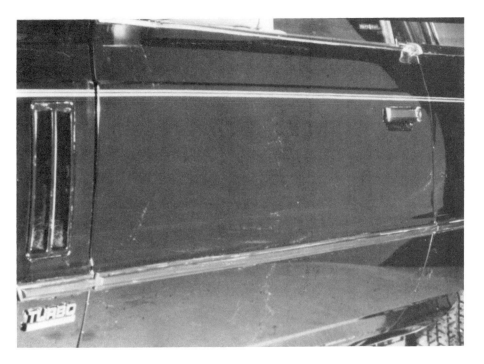

FIGURE 8.11   Diffracto sight image of body side (without).

edge radius.   All of these variables can be determined using
machine vision, but very few installations are known to exist
even though such techniques are must faster than touch trigger probes, for example.

The inspection of cutting tools can be done in many locations, e.g., in a presetting area, magazine, or on a spindle,
depending on the time allowed and the immediacy of data required.   Information can be used to feed back data for machine
offsets or tool setting or simply to determine whether or not to
change a tool.

In-process dimensional checks can also be carried out to
feed back data to the tool.   One example provided by our company was for the tip grinding of rotating turbine blades, where
shutters on the sensor unit are opened up, measurements are
taken, and the shutters are closed back down to allow the
grinder to equalize the tip dimension of all blades in the stage.

In general, all types of vision sensors can be incorporated into tools if miniaturization and other prerequisites are complete. The foremost question is whether the environmental aspects of particular coolant flow, oil mist, and chips can be dispensed with at least during the time at which the measurement is taken. It is noted that when operating in a noncontact manner, the part itself should, in general, have a homogeneous coolant film or no coolant at all. No big droplets are generally allowed if one is to make good measurements, although there are ways of getting around this in some cases.

FUTURE THRUSTS AND CHALLENGES

Clearly, a major thrust is to move machine vision-based sensing, and indeed sensing of every sort, closer to the machines to provide the ultimate in direct feedback to the process with the least scrap to obtain the closest control. In this area, substantial further work needs to be done to determine where such sensors can be incorporated and with what precautions they can be made to work.

Increasing Cost-Effectiveness

As a general trend, we can expect by the end of the decade to see substantial cost/performance improvements in machine vision sensing. At the high end, this will result in nearly total replacement of human visual inspection and reliable robotic guidance in large numbers of material handling and assembly applications.

New Systems

Besides the clear-cut cases of parts inspection and inspection of tools, there are areas which can lead to new kinds of machining or forming systems. For example, one of the features of vision-type sensing is the ability to take a great deal of data quickly. This can allow optimization of casting location in machining stations, modification of feed rates, and the like. Contoured surfaces can also be made very quickly and their shape fed back to the machines particularly of interest—for example, on surfaces such as combustion chambers, gears, turbine airfoils, and dies. This type of feedback can lead to a much more optimized machining process in relation to the function of the part.

There is also a potential for specialized sensors to be built for machine tool feedback. For example, one might automatically determine the contour, volume, or other combustion chamber characteristics in engine cylinders and feed back data to one or more machining operations to build essentially "perfect" engines. Such thinking might also be applied to gear manufacture, where the comprehensive gear tooth surface data could be used to trim gears to fit.

Finally, it is clear that all automation around the machine tools—robots, automated guided vehicle systems (AGVS), etc.— will increasingly employ machine vision to relieve part positioning requirements and to provide the intelligent inputs required to allow the system to deal with a myriad of other variables.

APPENDIX A: ADVANTAGES OF MACHINE VISION

Dimensional Gauging and Inspection Advantages

    1. Noncontact with large standoff:

- Freedom from crashes, wear, and associated maintenance.
- No servo actuation required.
- No part or probe damage.
- Ease of implementation in process.

    2. Speed:

- More parts/hour.
- Fewer gauges.
- More checks per part.

The speed advantage over conventional gauges can run from 1.4 times (crankshafts) to 5 times (bolts).

    3. Accuracy:

- All-digital, driftless, and linear.
- Freedom from wear or contamination due to contact with parts.
- Requires only one master, and in some cases it is a factory-preset calibration.
- No part distortion error.
- No contact bounce on moving or rotating parts.

4. Large range:

- More parts can go through same gauge.
- More flexibility for part changeover.

5. Consistency, vis-a-vis human inspector:

- True 100% inspection rather than 60 to 85% maximum.
- Same objective criteria day in, day out.

6. Reliability of inspection: Good machine vision systems seldom quit or drift. This is due primarily to the noncontact operation and all-digital sensing and processing.

7. Capable of performing checks in difficult places and situations, e.g., for snap ring grooves, small holes, recesses, and threaded holes and in unusual environments.

8. Immediacy of feedback: Faster data gathering means faster data availability for process correction. The vision sensor, properly employed, is unexcelled for providing instant, reliable information on which further decisions can be confidentially based.

Robot and Machine Guidance Advantages

Many of the same advantages apply, with some additional ones related to machines:

1. Reduced fixturing or handling requirements using vision offsets based on part features (this can result in large cost savings).

2. Increased ability to deal with out-of-specification parts.

3. Tighter control of operations than could be provided by conventional fixturing.

APPENDIX B: ILLUSTRATIONS OF MACHINE
VISION TECHNOLOGIES

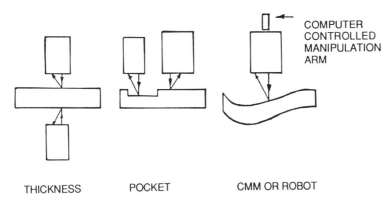

THICKNESS   POCKET   CMM OR ROBOT

FIGURE 8-B1 Laser triangulation sensors (laserprobe): Typical applications.

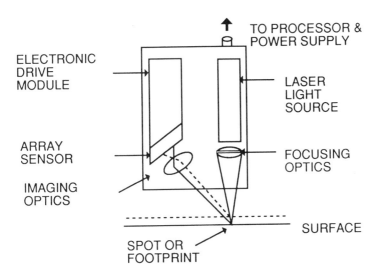

FIGURE 8-B2 Laser triangulation sensor (laserprobe):
Operation

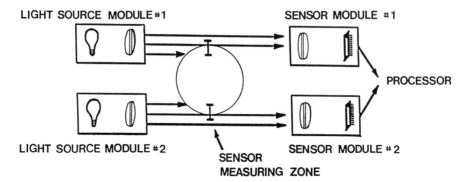

FIGURE 8-B3   Dimensional measurement sensor (DMS):   Diameter or length of large part (differential operation).

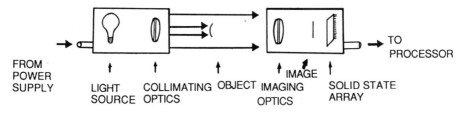

FIGURE 8-B4   Dimensional measurement sensor (DMS):   Operation.

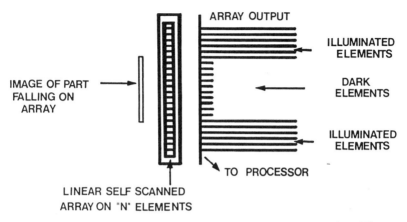

FIGURE 8-B5   Dimensional measurement sensor (DMS):   Linear array operation.

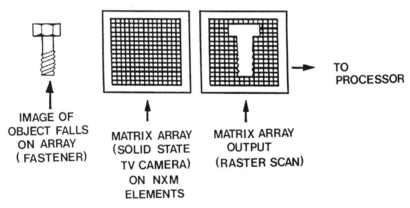

FIGURE 8-B6   Dimensional measurement sensor (DMS):   Matrix array operation.

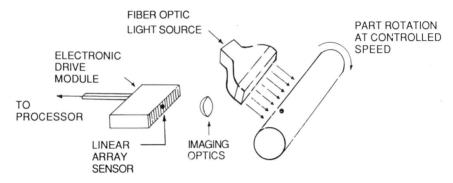

FIGURE 8-B7   Surfscan external surface flaws (SF):   Operation.

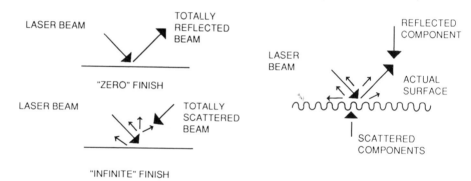

FIGURE 8-B8   Flawsensor (SF and BF):   Signal analysis.

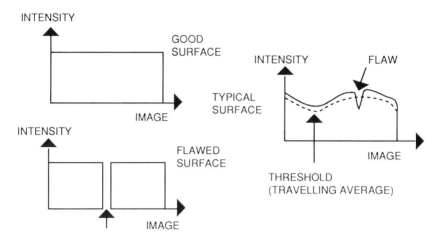

IDEAL SURFACES
ARRAY SCAN

DEFECT-"BLACKHOLE"

FIGURE 8-B9   Laser surface finish (lasersurf):   Operation

9
Fiber Optic Probes for In-Process Control

CURTIS D. KISSINGER / Mechanical Technology, Incorporated, Latham, New York

INTRODUCTION

The commercial development of efficient and cost-effective optical fibers and peripheral components for the telecommunications industry during the past decade has spawned a new generation of physical measurement transducers and sensing systems which utilize light energy as the primary signal transmission, conversion, and detection mechanism. This powerful combination of light energy and the means to harness it offers experimentalists and manufacturers a unique opportunity to create and market measurement and control devices having features and capabilities which were beyond the reach of previous transducer technologies.

Although we still know relatively little about the true nature of light, we do know a considerable amount about its characteristics and how to utilize its various properties for our purposes. This chapter presents a unique gauging sensor utilizing fiber optics which lends itself to both on-line and off-line measurement applications.

TECHNIQUES

Fiber optic transducers generally fall into one of four categories:

1. Intensity-modulated devices
2. Interferometric devices
3. Polarization-based devices
4. Wavelength-modulated devices

A summary of these techniques can be found in Refs. 1, 2. Since most in-process measurement and control applications require ease of installation, low cost, and reliability, the majority of fiber optic systems in industrial use today employ intensity-modulated devices because their inherent simplicity meets the needs of most industrial applications and their performance often rivals that of the more complex techniques. Manufacturers of intensity-modulated fiber optic sensors have devised numerous clever and practical means to adapt their equipment to a wide variety of experimental and commercial applications. Intensity-modulated devices usually fall into one of two primary categories: reflective and transmissive. Both of these methods can be used in either the on-off mode or a proportional mode. The on-off mode is employed for counting or determining the absence or presence of parts in a great variety of commercial applications, whereas the proportional mode is generally the more useful for in-process measurement and control since it gives a relative output signal which can be interpreted as a measurement of some physical property such as position displacement or a dimension.

THE FIBER OPTIC LEVER TRANSDUCER

Most proportional systems in use today employ some form of a derivative of the fiber optic lever displacement transducer. A complete theoretical analysis of the principle of operation of this transducer has been provided by Cook and Hamm (Ref. 2). This analysis is highly recommended for readers who may want to delve more deeply into this concept.

Fiber optic lever displacement sensors utilize adjacent pairs of light-transmitting and light-receiving fibers. The basis for the operating mechanism is the interaction between the source fibers and the field of view of the receiving, or detector, fibers. A greatly simplified illustration of this effect is shown in Figure 9.1.

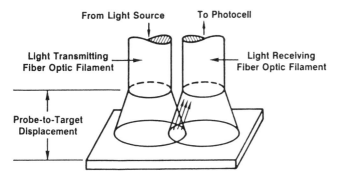

FIGURE 9.1   Displacement sensing mechanisms of adjacent fiber optic elements.

At contact, or zero gap, most of the light exiting the source fibers is reflected directly back into the same fibers, providing very little light to the receiving fibers and hence producing zero output signal.   An increase in the probe-to-target distance will result in some reflected light being captured by the receiving fiber.   This "increased distance-more light captured" relationship will continue until the entire face of the receiving fiber is illuminated with reflected light.   Further increases in distance will cause the diverging field of reflected light to exceed the field of view of the receiving fiber, thus causing a reversal in the output-versus-distance signal relationship.   The typical output-versus-distance response curve is shown in Figure 9.2.

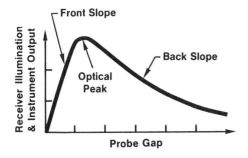

FIGURE 9.2   Typical fiber optic sensor calibration curve.

Also shown in Figure 9.2 are the three key segments of
the characteristic curve called the "front slope," the "back
slope," and the "optical peak." The front slope is the initial,
highly sensitive positive slope response extending from direct
contact out to the peak or maximum receiver output. The back
slope is the less sensitive, negative slope of the response extending from the peak on out to larger gaps. The optical peak
is the zero slope portion of the response curve, where the receiver fibers capture the maximum reflected energy relative to
the surface reflectance, and optical path attenuation. The front
slope response characteristic is a function of the fiber diameter,
the relative position of the transmit and receive fibers, and the
numerical aperture of the fibers. The back slope response
characteristic is primarily a field intensity inverse square law
function. A more complete description and theoretical analysis
of these functions is given by Cook and Hamm.

PROBE CONFIGURATIONS

Figure 9.3 shows the output-versus-distance response curves
of three different fiber relationships or arrays: the random
(R), or thoroughly intermixed transmitting and receiving fibers;
the hemispherical (H), or semicircular fiber grouping; and the
concentric transmit inside (CTI) fiber grouping. The fibers
used in these particular arrays were of the step-index type
having a core diameter of 0.0025 in., an OD of 0.0027 in., and
a numerical aperture of 0.63. Each array consisted of approximately 900 fibers contained in a stainless steel shaft having a
0.125-in. OD and a 0.086-in. ID. The R-type array gives the

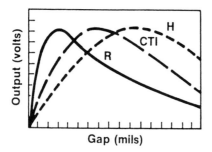

FIGURE 9.3 Output vs. distance for R, H, and CTI probe arrays.

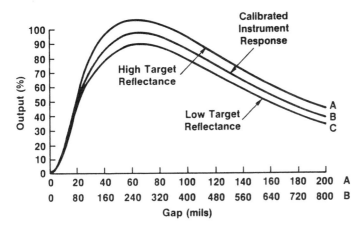

FIGURE 9.4 Output vs. distance for changes in reflectance.

most sensitive front slope and smallest sensing range; the H-type array gives the least sensitive front slope and greatest range; and the CTI array gives an intermediate response. The back slope sensitivity-range relationship follows a similar pattern for all three types. The output level at each optical peak was normalized for the purpose of comparison.

If source intensity is held constant and surface reflectance is changed, a family of curves, as shown in Figure 9.4, will be obtained for a 0.125-in. R-type probe. The form of the response curve remains essentially independent of surface color or gray-scale changes. If the source intensity is altered so that the magnitude of received light at the optical peak is the same regardless of reflectance, the response curves remain consistent, thus creating a sensing system having repeatable calibration factors.

If the geometry of the reflecting surface is changed from a specular to a diffuse nature, a slight change will occur in the back slope values even after the optical peak magnitudes are normalized. This effect is illustrated in Figure 9.5, which shows two response curves for an R-type probe calibrated to a 2-μin. AA surface (specular) and to a 32-μin. AA surface diffuse.

Table 9.1 shows some typical values of front and back slope range and sensitivity of a variety of probe configurations and

TABLE 9.1 Typical Fiber Optic Lever Characteristics

Probe number	Probe tip diameter (in.)		Front slope characteristics		Back slope characteristics		Optical peak (mils), midpoint
	Total	Active	Sensitivity (μin./mV)	Linear range ±5% (mils)	Sensitivity (μin./mV)	Linear range ±5% (mils)	
125R	0.125	0.086	0.6	4	18	70	18
125H	0.125	0.086	7.0	30	30	150	120
125CTI	0.125	0.086	3.5	15	25	80	75
062R	0.063	0.047	0.6	3.5	10	35	16
062H	0.063	0.047	3.0	14	12	55	50
047R	0.047	0.027	0.6	3.5	8	30	14
032R	0.032	0.019	0.55	3	5	25	12
020R	0.020	0.007	0.5	3	2.5	15	10

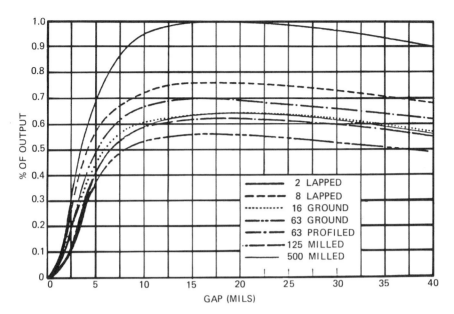

FIGURE 9.5 Output vs. distance for changes in surface finish.

sizes. The front slope range of R-type probe is basically independent of bundle diameter because a random probe operating
on the front slope is essentially a single-transmit, single-receive
configuration independent of the number of fibers used. However, the back slope range is proportional to overall diameter.
The front slope of an H-type probe is directly proportional to
the bundle diameter. In these probes, the entire semicircular
group of either transmit or receive fibers acts as a large, single fiber composed of many small fibers. Therefore, a larger
gap is required before the optical peak is reached.

TARGET MATERIALS

When target materials such as glass, plastic ceramic, or liquids
are used, some light will be absorbed into the material itself.
A positive offset will occur in the output signal at zero gap,
which will result in some loss of front slope range. Also, media
other than air in the sensing gap will alter the calibration curves.
Therefore, in either situation, a calibration must be taken under
actual or simulated operating conditions. Figure 9.6 shows an

FIBER OPTIC LEVER DISPLACEMENT TRANSDUCER
(0.0025 in., 0.62 N.A. Fibers; 0.086-in.-dia Bundle Size)

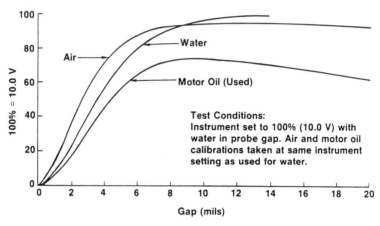

FIGURE 9.6   Change in gap calibration for different transmission media.

example which illustrates the change in calibration when the gap medium is changed from water, to air, to motor oil.

OPERATIONAL VARIATIONS

An interesting and useful variation of the basic fiber optic lever is obtained when a focusing lens system is placed near the sensing tip of the fiber optic probe (Ref. 3). The results for one such lens and fiber optic probe combination are shown in Figure 9.7. As can be seen, two peaks are now present with a very sharp null point midway between the peaks. The positive-going slope to the right of the null point is essentially a repeat of the front slope of a fiber optic probe operating without the lens system. Most important, however, the operating gap, or standoff distance, is approximately 100 times greater than without the lens system. The area to the left of the null point is essentially a mirror image of the right-hand portion.

Another interesting operating variation is obtained when the target movement is perpendicular to the axis of the probe. In this situation, the probe gap is usually set to correspond to the

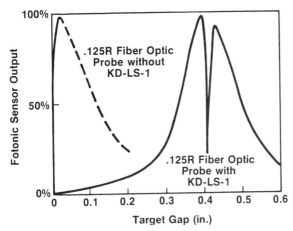

FIGURE 9.7   Fiber optic probe with lens system characteristics.

midpoint of the optical peaks to minimize any gap change effects. Figure 9.8 shows the output characteristics of a 0.125-in.-diameter H-type probe sensing edge motion parallel to the transmit/receive interface.   This calibration was taken with the probe

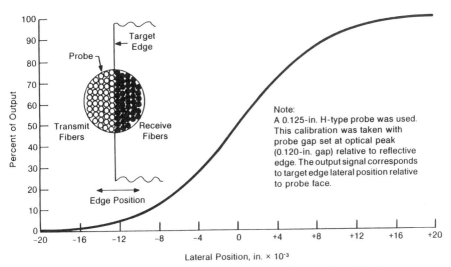

Figure 9.8   Output vs. lateral edge position—edge 90 degrees to transmit/receive interface.

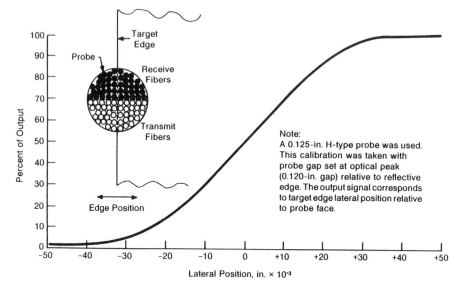

FIGURE 9.9 Output vs. lateral position—edge parallel to transmit/receive interface.

gap set at the optical peak (0.120-in. gap) relative to the reflective edge. The output signal corresponds to the target edge lateral position relative to the probe face. Figure 9.9 shows the same probe sensing edge motion perpendicular to the transmit/ receive interface. When an edge is not present, the same effect can be obtained by either attaching a stick-on target material or painting a reflective-to-nonreflective transition directly on the part being measured.

REFLECTANCE COMPENSATION

The fiber optic lever systems described up to this point can operate quite accurately as long as the reflectance of the target surface remains the same during the test period as during calibration. Although this condition can often be achieved, many applications exist where this is impossible or impractical due to remote or difficult probe locations or other secondary effects,

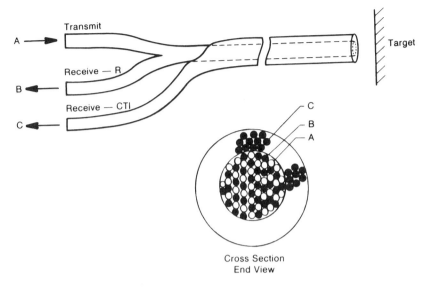

Cross Section
End View

FIGURE 9.10 Reflectance-compensated fiber optic probe with dual receivers.

such as time and/or temperature, which may alter the surface reflectance to an unknown degree during the operating period.

In an effort to overcome this problem, an improved fiber optic lever displacement sensing system has been developed. This system uses probes with a dual set of receive fibers having different optical levers. Since both optical levers are affected to the same degree by reflectance, an electronic circuit employing a ratio amplifier can be used to cancel the reflectance effect and to provide an output signal whose level is proportional to gap only. One such configuration is shown in Figure 9.10. This device employs a random transmit and receive core array surrounded by a coaxial group of receive-only fibers. The random core is about 0.050 in. OD and the CTI coaxial group is about 0.090 in. OD. The outer diameter of the steel tube sheath is 0.125 in.

Figure 9.11 shows the response curves of the R and CTI receivers and of the output of the ratio circuit. Tests have shown that the ratio output remains constant to better than ±1% of full scale over as much as an 80% drop in reflectivity, thus

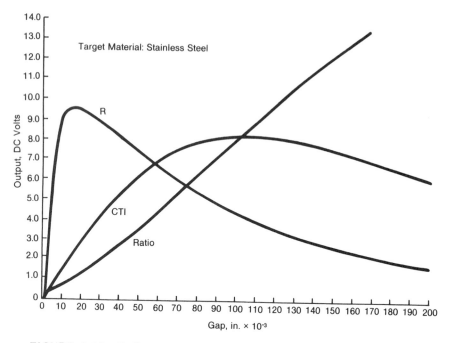

FIGURE 9.11   Reflectance-compensated Fotonic sensor—output vs. distance.

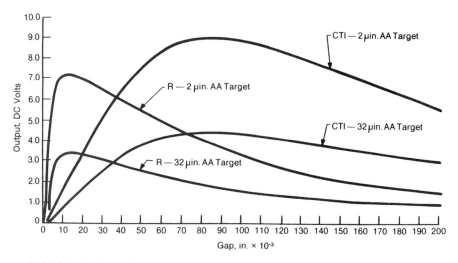

FIGURE 9.12   Reflectance-compensated Fotonic sensor—output vs. distance as a function of surface finish.

290

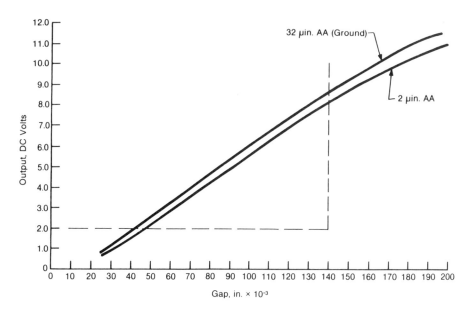

FIGURE 9.13 Reflectance-compensated Fotonic sensor—ratio output vs. distance as a function of surface finish.

thus providing accurate gap information over greatly varying degrees of surface reflectance.

Figure 9.12 shows the R and CTI signals obtained by using targets with a 2-μin. and a 32-μin. AA surface finish. Figure 9.13 shows the ratioed output with the same targets. A comparison of the two figures shows that the 1R and CTI signals have dropped by about 50%, whereas the slope of the ratio has changed by only about 2% and the position indication has changed by only approximately 5%. The ratio has changed because the diffuse surface of the 32-μin. AA finish reflects the light differently than the specular surface of the 2-μin. AA finish, which upsets the ratio signal slightly. An in-depth theoretical analysis concerning the aspects of the phenomena discussed in this paragraph is provided by Hoogenboom et al. (Ref. 4).

APPLICATIONS

Vibration displacement measurement of lightweight structures such as the magnetic head and flexure assembly (Refs. 5, 13) used

in hard disk drives is a good example of an application where
the inherent characteristics of the fiber optic lever displacement
transducer are essential in order that valid test results can be
obtained. Any additional mass or direct contact with these tiny,
flexible parts, even though their resonant frequency may not
exceed a bandwidth of a few kilohertz, would seriously alter the
actual operating state of the mechanism and thus give erroneous
information. In addition, the measurement of the dynamic behavior of the head position and support structure requires the
same characteristics from the measurement device.

Experimental and production measurement of the vibration
displacement, mode, and wave shape of the high-frequency motions present in ultrasonic welding equipment (Ref. 6) used in
bonding the tiny connecting wires used to manufacture integrated circuits can also be done with the fiber optic lever displacement transducer. In this application, the resonant frequencies
are often as high as 50 kHz; thus, even extremely small displacements can generate forces many thousands of times the
force of gravity. This prohibits the use of additional mass
loading as well as the ultralightweight structures previously
mentioned.

Rolling element bearing performance and health (Refs. 7, 8)
can be monitored by installing a fiber optic lever displacement
transducer in machinery to sense the minute radial displacements
in the surface of the outer race as each ball or roller passes the
transducer location. To do this, a small access hole is drilled
in the machine structure to permit the sensing tip of the transducer to view the outer bearing race. The transducer is referenced to the body or casing of the machine itself, thereby can-
celing extraneous vibration displacements such as one-per-revo-
lution imbalance or machinery mounting effects and thus permitting motion measurements in the order of microinches to be made
in the presence of background vibrations of much greater amplitude. It has been found experimentally that this technique is
often a better early warning indicator of incipient bearing failure (Ref. 8) than case-mounted accelerometers.

The essentially zero hysteresis operation of fiber optic lever
displacement transducers makes them well suited for applications
where repeatability of mechanical devices must be measured very
precisely. The measurement of the positioning repeatability of
the head position mechanism used in disk drives, as previously
mentioned, is one such application. Another application where

similar measurements must be made is in the study of the repeatability of mechanical positioning mechanisms, such as lead
screws, used in precision machine tools (Ref. 9).

The fiber optic lever displacement transducer, when combined with a lens imaging system (Ref. 3) which gives a conver-
gent focal point forward of the sensor face as opposed to the
divergent field of the fiber optic lever displacement transducer
used without a lens, offers additional unique characteristics
which can be utilized in several ways. One such device makes
use of the null which occurs at the focal point for the measurement of a variety of materials whose reflectances and physical
characteristics vary widely, without incurring errors due to
these variations. A servomechanism is used to lock onto the
null point and track its position change and thus produce a
measurement essentially independent of reflectance (Ref. 10).
The fiber optic probe and lens combination is used in high-
volume production calibration of clinical thermometers. In this
application, the divergent field permits the sensor to be focused
at the subsurface interface of the glass body and the tempera-
ture-indicating fluid, usually mercury or alcohol with a dye additive. This application also makes use of the lateral sensing
mode of operation as illustrated in Figures 9.8 and 9.9.

The fiber optic lever displacement transducer described
herein continues to find new and unique applications because
of its noncontacting nature (that is, it imposes no load on the
device being measured), because it has a high-frequency response, and because of its small size coupled with inherent
simplicity and operational practicality.

REFERENCES

1. M. E. Patton, "Fiber Displacement Sensors for Metrology
    and Control," Optical Engineering, Vol. 24, No. 6, Novem-
    ber/December 1985.
2. R. O. Cook and C. W. Hamm, "Fiber Optic Lever Displace-
    ment Transducer," Applied Optics, Vol. 18, p. 3230,
    October 1, 1979.
3. C. D. Kissinger and R. C. Smith, "Improved Noncontact
    Fiber Optics/Lens Displacement Measuring System," Proceed-
    ings of the Technical Programs, Electro-Optics Design Con-
    ference, pp. 372-377, 1973.

4. L. Hoogenboom, G. H. Allen, and S. Wang, "Theoretical and Experimental Analysis of a Fiber Optic Proximity Probe," Proceedings of the International Society of Optical Engineers (SPIE), Vol. 478, Fiber Optic and Laser Sensors 11, May 1-2, 1984.

5. M. E. Patton, and M. W. Trethewey, "A Technique for Non-Intrusive Modal Analysis of Ultralightweight Structures," 5th International Modal Analysis Conference, Imperial College of Science and Technology, London, April 6-9, 1987.

6. F. J. Crispi, G. D. Maling, Jr., and A. W. Rzant, "Monitoring Microinch Displacements in Ultrasonic Welding Equipment," IBM Journal of Research and Development, Vol. 10, No. 3, May 1977.

7. G. J. Phillips, "Bearing Performance Investigation Through Speed Ratio Measurements," 23rd Annual Meeting of the American Society of Lubrication Engineers, April 17-20, 1978.

8. M. E. Hampson, J. J. Collins, M. R. Randall, and S. Barkhoudarin, "SSME (Space Shuttle Main Engine) Bearing Health Monitoring Using a Fiber Optic Deflectometer," Proceedings of Advanced Earth to Orbit Propulsion Technology Conference, Huntsville, Ala., May 1986.

9. W. R. Moore, "Repeatability of .2 Microinches!" American Machinist Magazine, August 29, 1966.

10. R. L. Jackson, "Micro Trak—A Precision Non-Contact Displacement Sensor," AutoFact Sensors '85 Conference, CASA of SME, Detroit, Mich.

11. C. W. Fetheroff, T. Derkacs, and I. M. Matay, "An Automated Plasma Spray Feasibility Study," NASA, Report (NASA CR-159579), Contract NAS3-20112, May 1979.

12. A. D. Kersey and A. Dandridge, "Fiber Optic Sensors for Industrial Use," Chemical Processing Magazine, November, 1989.

13. J. Castagna, "A Method of Determining the Modes of Vibration of Disk Drive Heads and Suspensions During Operation," Sound and Vibration Magazine, August 1989.

10
Quality and Automated Process Gauging

THOMAS H. GEORGE / TRW Inc., Cleveland, Ohio

INTRODUCTION

Customers do not knowingly and willingly pay the same price for the same product features and second-best quality. The apparent exceptions—purchase selection dictated by a sense of duty ("Union Made" and "Made in the USA") or by expediency (the need is urgent and only the second best is immediately available)—only reemphasize the point. Compelling external reasons are required to persuade a customer to settle for less than the best-quality product available at the price. This is the bedrock principle that all manufacturing organizations must live by and that must govern every decision regarding product and process design and every stage of production. Numerous studies have confirmed the principle that a reputation for best product quality is the surest route to company growth in market share and profitability (Ref. 1). Such reputations are earned only as customers learn by experience which products are consistently delivered defect free and consistently perform their stated functions exactly and reliably.

Gauging is intended to improve product quality, but it also increases the product cost. This must always be a design consideration. The willingness of a potential customer to purchase a given product is severely bounded by competitive price. The

customer may be willing to forgo a minor function of a product
in favor of a competing product lacking this function but otherwise of clearly superior quality. The customer is not likely to
give up a major function. For this reason, superior product
quality rarely commands more than a modest price premium; too
large a price premium brings a product into competition with
products that have significantly enhanced or expanded functions.
Therefore automated process gauging, like manual inspection,
cannot be used as a brute-force method of improving quality;
it must be applied sparingly and selectively as needed to guarantee the best quality at the lowest possible cost.

A product designer must make trade-offs between product
functions, performance, reliability, and cost. Truly creative
design also includes design for manufacturability based on a
thorough understanding of the available or attainable manufacturing processes and their statistically determined process capabilities. Only by integrating the considerations of process selection and control into the total design process can the objective of maximum product functions, performance, and reliability
at the most competitive price be fully realized. The implication
here is that, to the maximum extent possible, process selection
will focus on thoroughly capable and stable processes which can
be kept under statistical control at lowest cost with minimal manual sampling and gauging. Automated process gauging will,
therefore, find its best applications in controlling state-of-the-
art (i.e., incapable) processes, marginally capable processes,
and potentially capable processes with high drift rates. Thus
our attention should be concentrated on such processes and
their implications.

There is another less obvious but extremely important consequence of the low-cost constraint. The trend to increasingly
customized products and short lead times has led to a revolution
in manufacturing methods referred to as just-in-time manufacturing methods and world competitive manufacturing (Refs. 2, 3,
4). A key tenet of this approach is the reduction of setup times
to facilitate small-lot production. Hall (Ref. 5) has noted that
the real challenge and opportunity of automation lies not in automating the process but rather in automating the setup. The
former is relatively straightforward, given today's technology;
the latter is the real key to cost-efficient production because it
accelerates the shift to just-in-time production methods. It follows that rapid setup and calibration is increasingly important.

There are examples of 10-minute machine setups rendered inef-
fective by a 1-hour setup of the in-process gauging. In a
just-in-time manufacturing environment the setup of an entire
production cell or line is the longest individual setup time. The
targets are setup times of less than 10 minutes and preferably
less than 1 minute. The successes of the method in improving
quality, reducing lead times, and reducing costs make it certain that quick setup requirements will grow in importance.

TOLERANCES AND CONTROL

Tolerances are set to ensure that the completed product will
perform consistently and reliably according to specification and
to ensure product uniformity. It is important to distinguish
between the tolerances needed to ensure that the product will
function and will continue to function reliably and the tolerances
which serve only to control product uniformity. For convenience,
the former will be referred to as function tolerances and the latter as uniformity tolerances. The basic dimensions to which
these tolerances are applied will be referred to as function specifications and uniformity specifications. Unless specifically
noted, it should be assumed that tolerances are symmetric, that
is, the lower specification limit is the basic dimension minus one
half the tolerance and the upper specification limit is the basic
dimension plus one half the tolerance.

   It is to be expected that function tolerances will be tightened as process capability permits and must be held by sorting
if no capable process is available. Whenever possible, uniformity tolerances should be relaxed if no capable process is available.

   An example of the treatment of a uniformity tolerance quickly brings out the salient points. The product is a large cloth-
ing button with a tolerance of 0.010 in. on the outer diameter.
A statistical study of the capability of the buttonmaking process
determined its six standard deviations capability to be 0.014
in.—in other words, if the process is held perfectly centered
and the variation is normally distributed, 3% of the product
would be expected to be out of tolerance. In this situation
there are four possible courses of action: ship defective product, get the tolerance relaxed, institute manual or automated
gauging and sorting, or improve the process capability. The
first course of action must always be regarded as unacceptable.

A reputation for quality is a valuable asset which can only be created slowly with careful, sustained effort but may be quickly and easily destroyed. The third and fourth courses of action will certainly increase product cost. Therefore the customer, a garment manufacturer, was first asked if it would be possible to relax the tolerance. The first reaction was amazement—the buttonholes were not being sewn to such tolerances and such a minor difference would be undetectable to the final customer. On investigation, it was determined that the tolerance had been set to ensure that variations in button diameter would not cause the automatic button feeder to jam. If such variations could cause jams, then the tolerance would be a properly set function tolerance. In fact, further study showed that variations of as much as 0.020 in. caused no difficulty. The tolerance was relaxed to this value.

This simple example emphasizes the importance of distinguishing between uniformity and function tolerances. In this particular case, money spent on sorting or improving process capability would have been a disservice, since ultimately the increased costs are passed on to the final customer or the firm is weakened by a noncompetitive cost structure. Indeed, in some instances it will not be possible to get uniformity tolerances relaxed and so it will be necessary to resort to sorting or to process improvement. The essential point is that, as long as the uniformity tolerance is held, the variation within the tolerance is absolutely immaterial to the final customer and may even be undetectable. This means that if a process for a uniformity tolerance can be kept under complete statistical control, the resulting product is, for practical purposes, perfect. If complete control is not possible, the same can be said for the result of 100% inspection and sorting, provided the inspection and sorting process is statistically capable. This is seldom true of manual inspection but is a real possibility with automated inspection and sorting.

In contrast, for function tolerances clipping the output distribution by sorting and allowing drift within the tolerance should always be regarded as unacceptable if it can reasonably be avoided. Recently, much attention has been given to the Taguchi loss function (Refs. 6, 7), which emphasizes the loss associated with any departure from the basic dimensions. In many cases the societal costs associated with this concept cannot be evaluated with sufficient accuracy to give meaningful quantitative results. It is nonetheless clear that when function and reliability are at stake, any departure from the basic dimensions is undesirable.

A quantitative example will make the extent of the problem clear. Given a normally distributed output for which four standard deviations equals the tolerance, a centered process will produce a 4.5% nonconforming product, half over the upper specification limit and half under the lower specification limit. If undersize product must be scrapped while oversize product can be reworked, there is a tendency to target the mean of the output off center so that virtually no scrap will be produced. In this case, if the mean is targeted to lie three standard deviations above the lower specification limit, only 0.135% of the output should be scrap. On the other side, though, the mean will lie just one standard deviation below the upper specification limit and 15.866% of the output will require rework. (Note the increase from 4.5% to 16.0% nonconforming product. This is the reason for assuming symmetric tolerances unless there are specific reasons for requiring nonsymmetric tolerances.) If all of this output is successfully reworked to bring it just within the upper specification limit, then 84% of the final product will be above the basic dimension and 50% will be a full standard deviation above the basic dimension. Just this situation was encountered in the production of forged aircraft turbine blades. Unfortunately, the basic dimension in question was a carefully chosen compromise to minimize fuel consumption without seriously affecting maximum thrust. The shift of 50% of the output a full standard deviation from this carefully chosen basic dimension appreciably increased the engines' fuel consumption. Of course, the process target was immediately corrected as soon as the consequences of the previous targeting were fully understood. Often the consequences are not so dramatic, but a quantitative understanding of the consequences of clipping an output distribution by gauging and sorting and the effects of changing the distribution within the tolerance limits is a must for critical features which affect product performance and reliability.

It might seem that these are issues of more concern to the process and quality engineers actually operating the process than to the designers of automated process gauging. In fact, there are at least two reasons for gauge designers to take specific interest above and beyond the general desirability of always trying to understand the customer's objectives and problems.

First, the best data on process capability under actual manufacturing floor conditions come from sampling the output over
a considerable interval of time, an interval in which all of the
changing conditions which affect process yield are likely to
have occurred. If automated process gauging is being used,
it should capture the appropriate information. Second, the
gauge controller will often generate the necessary adjustment
signals for the process controller, and current technology
makes it possible to implement control strategies keyed to process characteristics.

For example, it is often assumed that the variation in the
output of a process will be normally distributed. For well-de-
veloped and capable processes this is generally a good assumption. The assignable causes of variation have been identified
and minimized or removed, so the remaining variation is due to
many minor causes which contribute roughly equally to the
overall effect. In such cases the central limit theorem applies
and the output distribution will very closely approximate the
normal distribution. But recall that the best applications of
automated process gauging should be to state-of-the-art and
marginally capable processes. For such processes a single or
a few assignable causes may be the source of most of the observed variation. In these cases the distribution will be characteristic of the assignable causes and need not even remotely
approximate a normal distribution. In addition, there are the
cases of measurements of departures from an ideal (flatness,
straightness, etc.), for which the distribution is also clearly
not normal. (Many of these may actually be cases of folded
normal distributions; Ref. 8.) When the distribution is not
normal, good sampling can permit estimation of the statistics
of a theoretical distribution which closely approximates the actual distribution of the process output. If such a theoretical
distribution and statistics are available, the microprocessor of
a gauge controller could include the capability to implement an
optimum process control strategy based on this information.

CONTROL STRATEGY

All automated in-process and postprocess gauging for process
control should generate adjustment signals based on sample statistics (mean and standard deviation). There are three reasons
for this unqualified statement:

1. Regardless of actual tolerances and process capabilities, the objective for all function characteristics is to hold the output variation to the absolute minimum possible and to keep the process as closely centered on the basic dimension as possible. The limit is always the underlying random variation in the process. Any attempt to control random variation results in instability and less than optimum performance. Control based on sample statistics assists in the detection of any systematic component of the variation.

2. The capture of sample statistics and data over extended periods of process time provides the best assessment of process capability and, if the output distribution is not normal, the information necessary to determine the best mathematical approximation of this distribution and to devise the most effective control strategies.

3. If no effort is to be made to determine the actual form of the output distribution, then control based on sample statistics has the best prospect of succeeding because, by the central limit theorem, the distribution of the means of the samples will tend toward the normal distribution even if the distribution being sampled is far from normal.

Accepting that all gauging and control will be based on sample statistics, four questions remain: how to determine the statistics to be used, what methods to use to detect out-of-adjustment and out-of-control conditions, how to estimate appropriate adjustments, and how best to combine out-of-adjustment signals and adjustment estimates. To answer these questions, several control and adjustment methods were tested against a set of normally distributed random numbers to which systematic variations were added. The next five subsections summarize the methods used and the fourth section presents the results of the tests.

Determination of the Standard Deviation
of the Process

An $\bar{X}$ and R or an $\bar{X}$ and $s_{\bar{X}}$ statistical process control chart provides excellent data for the estimation of the true standard deviation of the process, σ (Refs. 9, 10). (Note: The estimate of σ is s_X, the standard deviation for variations of individual observations, X. The control chart is usually plotted with the means, $\bar{X}$, of samples of size n. The standard deviation for

variations of observations of these means, $\bar{X}$, is $s_{\bar{X}} = s_X / \sqrt{n}$.
Control charts for individual observations are also possible and
will be used in this discussion. Considerable care has been
used to apply the notation consistently but the reader must always keep the context in mind, that is, whether the discussion
refers to individual measurements or the means of samples of
size n.)

If sample data are accumulated at intervals over a period
of time and the control chart indicates that the process was
under control in both $\bar{X}$ and R or in both $\bar{X}$ and $s_{\bar{X}}$, then the
accumulated data may be used for a capability study to provide
a good test of the normality of the output distribution and a
good estimate of the standard deviation of the process. Provided the under-control conditions are met, it is perfectly per-
missible to make process adjustments in the intervals between
samples and use the resulting data for the capability study.
Thus in most cases it is possible to get good estimates of process capability under extended shop floor conditions.

This procedure occasionally leads to an overestimate of the
process capability. In these instances an analysis of variance
calculation of the standard deviation between groups will be a
better estimate of the real process capability. If this approach
is used, either no process adjustments are made or process adjustments become a component of the process variability. The
necessary calculations for this special case are given below.

Since the CPU of the gauge controller is to do most of the
work, there is no reason not to go first class, that is, to pre-
fer the X and s_X chart. Accordingly, the standard deviation,
s_i, of each sample should be calculated from the formula

$$s_i = \sqrt{\frac{\sum_{j=1}^{n} (x_{ij} - \bar{X}_i)^2}{n - 1}} \quad \text{where} \quad \bar{X}_i = \frac{1}{n} \sum_{j=1}^{n} x_{ij} \qquad (10.1)$$

where x_{ij} is the jth observation in the ith sample and n is the
sample size, which is assumed to be fixed. (This and subsequent formulas can readily be generalized to the case of varying sample sizes, but such a situation would rarely be appropriate for process controllers.) After k such samples have
been collected (in general, k should be at least 20), a pooled
standard deviation, s_p, is calculated from the formula

$$s_p = \sqrt{\frac{1}{k} \sum_{j=1}^{k} s_i^2}$$ (10.2)

Assuming the process is under control, this pooled standard deviation will be used as the estimate of the true standard deviation, σ, of the process.

[Special case: In most cases the pooled standard deviation, s_p, is an excellent estimate, but occasionally it significantly underestimates σ. This is easy to check, however. If the process is under control, combine all the data and calculate a standard deviation, s_X, from the formula

$$s_X = \sqrt{\frac{1}{nk - 1} \sum_{i=1}^{k} \sum_{j=1}^{n} (x_{ij} - \bar{\bar{x}})^2} \quad \text{where} \quad \bar{\bar{x}} = \frac{1}{nk} \sum_{i=1}^{k} \sum_{j=1}^{n} x_{ij}$$

(10.3)

In effect, a classic analysis of variance experiment has been performed by taking k samples of size n from an ongoing process (Ref. 11). A parametric model would test the proposition that there is no significant difference between the means of the various samples ($\bar{X}$ under control). Since $\bar{X}$ is under control, a component of variance model is more appropriate to test the proposition that there is no significant difference between the variances of the various samples (s_i under control). We believe we have already tested this point with the control charts, but there is an exception. In the analysis of variance it is shown that the sum of squares in equation (10.3) can be resolved into two terms

$$\sum_{i=1}^{k} \sum_{j=1}^{n} (x_{ij} - \bar{\bar{x}})^2 = \sum_{i=1}^{k} \sum_{j=1}^{n} (x_{ij} - \bar{x}_i)^2 + \sum_{i=1}^{k} \sum_{j=1}^{n} (\bar{x}_i - \bar{\bar{x}})^2$$

(10.4)

If we define the between-groups standard deviation, s_b, by the formula

$$s_b = \sqrt{\frac{1}{k - 1} \sum_{i=1}^{k} \sum_{j=1}^{n} (\bar{x}_i - \bar{\bar{x}})^2}$$ (10.5)

then equation (10.4) can be written as

$$(nk - 1)s_x^2 = k(n - 1)s_p^2 + (k - 1)s_b^2 \qquad (10.6)$$

If, in fact, s_p and s_b are equal, then s_x and s_p are equal.
In practice, these numbers will not be exactly equal. An F
test with $n - 1$, $nk - 1$ degrees of freedom can be used to
test whether the ratio s_b^2/s_p^2 is significantly greater than one.

Why should this occur? It may happen that the full variability is not seen in any short interval for a variety of reasons,
for example, material variations, changes in temperature, humidity, hydraulic oil temperature, or automated process adjustments. Strictly speaking, these are assignable causes which
should be identified and eliminated. This is not always immediately possible. In such instances between-groups standard
deviation, s_b, is the appropriate value to use when calculating
process capability and control limits for $\bar{X}$. The pooled or
within-groups standard devication, s_p, is still the appropriate
value to use when calculating control limits for R or $s_{\bar{X}}$.]

Definition of Process Capability

The basic concept is that, when sampling from a normal distribution, the probability of a value lying more than three standard deviations above the mean of the distribution is fairly
low—indeed, is 0.135% for a perfect normal distribution. Since
the value could lie either above or below the mean, the total
probability of observing a value three or more standard deviations from the mean is 0.270%. Note that for a production run
of 1 million pieces this would amount to 2700 values three or
more standard deviations from the mean. If the specification
limits are placed at $\pm 3s_X$ from the basic dimension and the process is kept perfectly centered, this equates to 2700 parts per
million (ppm) nonconforming product. For a function specification this is an unacceptably high defect level. Therefore, it is
usually held that the specification limits must lie at least $\pm 4s_X$
from the basic dimension for the process to be considered capable. The $\pm 3s_X$ or, equivalently, the six standard deviation interval has long been used in discussions of process capability.
Thus the definitions of capability are based on this value.

The potential process capability, c_p, is defined as

$$c_p = \frac{USL - LSL}{6s_X} \qquad (10.7)$$

where USL is the upper specification limit, LSL is the lower specification limit, and s is the standard deviation of the process. The actual process capability, c_{pk}, is defined as

$$c_{pk} = \frac{\text{lesser of (USL - basic dim.) and (basic dim. - LSL)}}{3s_X}$$

(10.8)

For a perfectly centered process the two are equal. Controllable processes should be held centered as closely as possible on the basic dimension.

For a function specification c_{pk} should be at least 1.33 for the process to be considered capable. Given a perfect normal distribution, this would ensure that no more than 64 ppm nonconforming product would be produced. The need for early detection and correction of systematic departures from the target value is clear. The need to avoid unnecessary corrections should be equally clear, since these will generally move the process mean further from the target value.

Gauge Capability

An absolutely essential but occasionally neglected aspect of process control is confirmation that the gauges in use are appropriate for the task. This goes beyond simply ensuring that the gauge calibrations are up-to-date and traceable to the appropriate standards; that only guarantees that the gauges are within their design specifications, not that they are appropriate to the task at hand. What are required are gauges which will not contribute significantly to the variability of the process. This is generally taken to mean that, at an absolute maximum, the standard deviation of the gauge in the range of interest must not exceed one fifth of the tolerance. This maximum should be approached only for extremely tight tolerances which force measurements at the limit of current gauge capability.

The required gauge capability is readily checked. A set of parts—at least 20—of the type to be gauged are obtained and each part is marked with an identifying number. Each part is then gauged several times (5 times is adequate), the results are recorded, and, when the measurements are complete, a standard deviation is calculated for the results for each part. These standard deviations are pooled to obtain the required standard deviation for the gauge. If, for any reason, the parts must be

gauged manually, it is best to conduct this as a blind experiment; i.e., before each round of measurements the order of the
parts is changed and the person performing the measurement is
not aware of the part identification.

A further point of concern arises when a family of gauges
is in use. For example, the same characteristic is being gauged
once as the part is produced and a second time with a second
gauge at final inspection. In this case it is possible that both
gauges are properly calibrated and capable but are not adequately correlated in the particular range of interest. If so, one
gauge will accept on the high side parts that the second gauge
will reject, and will reject on the low side parts that the second
gauge would have accepted. This possibility is also readily investigated. The capability study is performed on each gauge.
In each case the standard deviation is within the requirement.
The standard deviations are then pooled, the mean result is
calculated for each gauge, and Student's t-test is used to determine whether the difference in these means is significant.
If it is, the gauges are not adequately correlated. If a large
family of gauges is in use and all are proved capable, the test
of the means should be applied to the two most widely differing
means. If the difference proves significant, one of the gauges
must be eliminated and the test repeated with the next two most
widely differing means until a satisfactorily correlated family is
identified. If one of the gauges is the customer's final accep-
tance gauge, it must remain a member of the set—gaining the
customer's agreement to include it in the study in the first
place can prevent many acrimonious arguments subsequently.

Detection of Out-of-Adjustment and
Out-of-Control Conditions

Given the selection of $\bar{X}$ and $s_{\bar{X}}$ data to determine the variability
of the process, the use of this method to detect out-of-adjust-
ment conditions might seem a foregone conclusion. This is not
the case. Another viable choice is the Cusum method. Furthermore, there are variations on both of these methods designed to
enhance the early detection of departures from the process mean
or target value. A subsequent section—Evaluating Combined
Out-of-Adjustment Tests and Adjustment Algorithms—will contrast the performance of these methods. This section defines
the four out-of-adjustment detection methods to be evaluated.
In all cases it is assumed that process adjustments (offset

adjustments) are possible to keep the process centered on the basic dimension, that it is desired to detect when these adjustments should be made, and that it is just as important to avoid making unnecessary adjustments. Additionally, indications of degradation of process capability should be detected.

Note: The term out-of-adjustment is not standard terminology but is used throughout this discussion to distinguish clearly between situations calling for a simple adjustment of an ongoing process and situations which require special action to determine whether a degradation of process capability has occurred.

The Basic $\bar{X}$ and $s_{\bar{X}}$ Control Chart

The basic $\bar{X}$ control chart is constructed by determining the estimated standard deviation of the population of $\bar{X}$ values, $s_{\bar{X}}$; using this estimated value to determine the control limits, $UCL_{\bar{X}}$ and $LCL_{\bar{X}}$, defined as the basic dimension $\pm 3s_{\bar{X}}$; and then plotting the observed values of X in sequence on a graph showing these limits. A solid horizontal line at the basic dimension, X', indicates the target value and dashed horizontal lines at $UCL_{\bar{X}}$ and $LCL_{\bar{X}}$ indicate the control limits. A typical chart is shown in Figure 10.1 with an out-of-adjustment observation circled.

Observed values falling within the control limits are judged to occur sufficiently frequently as random variation that no process adjustment is required. This avoids overcontrol. Observations falling outside the control limits occur only infrequently as as random variation (0.0027 for the normal distribution) and are

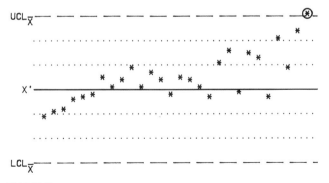

FIGURE 10.1 Basic X chart showing one point outside the upper control limit.

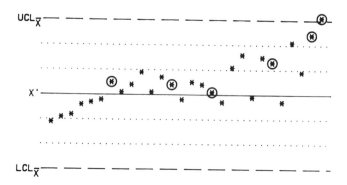

The comprehensive $\bar{X}$ chart. The circled points mark out-of-adjustment signals in the following sequence: Trend of seven, seven of seven on one side of the mean, ten of eleven on one side of the mean, four of five in the one to three sigma band, two of three in the two to three sigma band, and one point outside the upper control limit.

FIGURE 10.2   Comprehensive X chart.

judged more likely to represent process drift. In these cases a process adjustment is to be made to recenter the process on the basic dimension.

The Comprehensive $\bar{X}$ and $s_{\bar{X}}$ Control Chart

In an effort to achieve the earliest possible detection of process shift from the basic dimension and yet avoid overcontrol in response to random variations, a number of additional out-of-adjustment signals have been developed. The thought is to identify events that are statistically unlikely while the process remains stable and centered on the basic dimension but that would be very likely to occur if the process drifts significantly away from the basic dimension. To describe and use these it is convenient to add dotted horizontal lines to the control chart placed at $X' \pm s_{\bar{X}}$ and at $X' \pm 2s_{\bar{X}}$. Some of the out-of-adjustment tests used and their probability of occurrence as random events when sampling from a stable, normally distributed population are as follows:

Two of three in the $2s_{\bar{X}}$ to $3s_{\bar{X}}$ band, P = 0.00270
Four of five in the $1s_{\bar{X}}$ to $3s_{\bar{X}}$ band, P = 0.00534
Seven of seven in the 0 to $3s_{\bar{X}}$ band, P = 0.01533

Ten of eleven in the 0 to $3s_{\bar{X}}$ band, P = 0.00569
Trend of seven increasing (or decreasing) observations, P =
    0.00039

Figure 10.2 illustrates these tests on a control chart.

Note that the calculated probabilities assume no observations
completely outside the $3s_{\bar{X}}$ limits (UCL$\bar{X}$ and LCL$\bar{X}$), that they
include both the high (above X') and low (below X') cases, and
that they are not mutually exclusive (not additive). In practice,
the occurrence of a two out of three out-of-adjustment signal
would call for an immediate process adjustment and thus would
preclude the observation of a four out of five out-of-adjustment
signal of which these were the first two or three points, etc.

A number of other out-of-adjustment tests have been developed and used, but there is clearly a limit to the number of
tests that can be used simultaneously without reintroducing the
problem which was to be avoided, namely overcontrol in response
to random variation. (For manual charting there is also a practical limit to the number of tests an operator can reasonably
apply.) Therefore, the above tests in addition to the test for
points completely outside the control limits can be reasonably
regarded as a comprehensive set of out-of-adjustment tests for $\bar{X}$.

The Standard Mask Cusum Chart

The standard Cusum chart offers an alternative method for early
detection of process drift from a target value, which, in this
discussion, will always be assumed to be the basic dimension as
it has already been emphasized that a controllable process should
almost always be centered on the basic dimension. The term
Cusum refers to the cumulative sum of the differences obtained
by subtracting the target value from each observed value. If a
process is stable and the observed values are symmetrically distributed about the target value, this cumulative sum should
wander about the target value, being sometimes positive and
sometimes negative but never departing far from the target value. Thus a discernible trend away from the target value would
represent an out-of-adjustment signal. Figure 10.3 shows an
example of a standard Cusum mask applied to a Cusum chart.

To make this test practical the Cusum is plotted as a function of the observation number, a mask is constructed representing the slope of the smallest unacceptable trend, and, as each
successive observation is plotted, the reference point of the

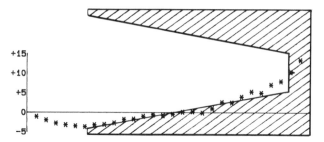

FIGURE 10.3   Standard Cusum mask.

mask is placed on the plotted point to determine whether a significant trend is present as indicated by the intersection of the
plotted Cusum line with the mask.   The exact shape of the mask
may be chosen to reflect control objectives and known features
of the process.   For the purposes of this discussion a standard
mask will be defined by the equation

$$M(j) = \pm(5 + 0.5 * j) * s_{\bar{X}} \qquad \text{for } j = 0 \text{ to } 19 \qquad (10.9)$$

The mask extends back from the current observation (j = 0) to
the nineteenth previous observation.   Since the zero of the mask
is centered on the current Cusum value, this latest value cannot
trigger an out-of-adjustment signal by itself but can do so only
by shifting the mask upward or downward relative to the Cusum
line to such an extent that one of the previously plotted points
lies on or outside the boundaries of the mask.

Two points should be noted about this mask.   The first is
that, given an absolutely stable process with no random variation but centered on a value displaced by Δ from the target
value, each successive observation will increment the Cusum by
Δ.   In this case if $19 * \Delta$ is less than $(5 + 0.5 * 19) * s_{\bar{X}}$, the
final point of the mask, no out-of-adjustment signal would ever
be generated and the Cusum would depart steadily from the target value.   In practice, some random variation is always present.
In the cases of interest it is significant compared to the anticipated short-term systematic variations and this random variation
will cause out-of-adjustment signals even in cases in which
$19 * \Delta$ is less than $(5 + 0.5 * 19) * s_{\bar{X}}$.   The second point is
that some systematic variations, e.g., a step shift, will require
a minimum mask length to guarantee detection in the absence of
any triggering due to random variation.   In fact, one of the

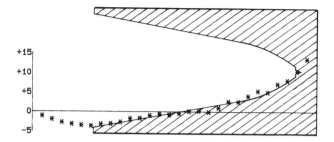

FIGURE 10.4   Semiparabolic Cusum mask.

adjustment algorithms to be considered requires a mask length of 14.   Such factors should be taken into consideration in the design of a mask for a specific process.

The Semiparabolic Mask Cusum Chart

Another factor for consideration in the design of a mask for a Cusum chart is the possible need for early detection of significant process shifts.   To accomplish this the mask may be closed in for the first few points.   To illustrate this, the following mask, referred to as a semiparabolic mask, is used in the evaluation of out-of-adjustment tests:

$$M(j) = \pm(1.25 + 2*j - 0.15 * j^2) * s_{\bar{X}} \quad \text{for } j = 0 \text{ to } 5$$

and $\hspace{10cm}$ (10.10)

$$M(j) = \pm(5 + 0.5*j) * s_{\bar{X}} \quad \text{for } j = 6 \text{ to } 19 \hspace{1cm} (10.11)$$

At $j = 1$ the mask boundary is now at $3.1 * s_{\bar{X}}$ instead of $5.5 * s_{\bar{X}}$.   If the process is perfectly centered and the Cusum is zero, the probability that the next observation will increment the Cusum by $3.1 * s_{\bar{X}}$ is 0.002 for samples from a normal distribution. This is sufficiently unlikely that a process adjustment is appropriate.   Figure 10.4 shows a Cusum chart and the semiparabolic mask.

The $s_{\bar{X}}$ Chart

The $s_{\bar{X}}$ chart is intended to detect loss of process capability. Out-of-adjustment signals from the preceding $\bar{X}$ and Cusum charts may also occur as a result of loss of process capability, and any extreme shift in $\bar{X}$ or Cusum should raise the possibility.

Since there is no routine adjustment for loss of process capability, out-of-control signals from the $s_{\bar{X}}$ chart must be given serious consideration.

The chart is constructed by plotting the observed values of $s_{\bar{X}}$ in the sequence of occurrence. By definition the standard deviation cannot be negative, so the scale of the abscissa will range from 0 to at least the upper control limit for $s_{\bar{X}}$ (UCLs$_{\bar{X}}$) to be placed three standard deviations of $s_{\bar{X}}$, $s_{s\bar{X}}$, above the mean value for $s_{\bar{X}}$. It is only here that the choice of an $\bar{X}$ and $s_{\bar{X}}$ chart introduces complexity. The first complication arises because $s_{\bar{X}}$ is not normally distributed (Ref. 12). As a result, it is most convenient to determine the upper control limit for $s_{\bar{X}}$ from tabulated values. The second complication arises because most of the published tables were developed in the 1930s for use in manual calculations. In consequence, the tables were developed on the assumption that the value of $s_{\bar{X}}$ would be determined by calculating the standard deviations of a number of small samples and averaging them together rather than calculating the square root of the pooled variance as described in the section titled Determination of the Standard Deviation of the Process. The pooled variance is an unbiased estimate of σ, the variance of the population, and is far more convenient in every other respect. This difficulty is resolved by using Table 10.1, which is calculated from the χ^2 distribution and using $s_{\bar{X}}/\sqrt{2n}$ as the estimate of $s_{s\bar{X}}$ (Ref. 13) or by referring to the published tables (Ref. 14). The last column, the probability of observing a value of $s_{\bar{X}}$ greater than UCLs$_{\bar{X}}$, is included because the definition of UCLs$_X$ as $s_X + 3 * s_{s\bar{X}}$, while consistent with the definition of the control limits for $\bar{X}$, does not imply the identical frequencies for observations outside these control limits.

Only observed values of $s_{\bar{X}}$ above the upper control limit for $s_{\bar{X}}$ will be used as out-of-control signals on the s chart. More comprehensive tests can be used, but such additional complexity is not generally needed. Of course, $s_{\bar{X}}$ "out-of-control" on the low side is of great interest since it indicates an improvement in process capability. This possibility should be checked from time to time and certainly after any modification of the process.

TABLE 10.1 Factors for Calculating Control Limits for an $s_{\bar{X}}$ Chart[a]

n	Central line	$LCLs_{\bar{X}}$	$UCLs_{\bar{X}}$	$P(s_{\bar{X}} > UCLs_{\bar{X}} \mid n)$
2	0.7979	0	2.2979	0.0216
3	0.8862	0	2.1110	0.0116
4	0.9213	0	1.9820	0.0082
5	0.9400	0	1.8887	0.0065
6	0.9515	0.0855	1.8176	0.0055
7	0.9594	0.1576	1.7612	0.0049
8	0.9650	0.2150	1.7150	0.0044
9	0.9693	0.2622	1.6764	0.0041
10	0.9727	0.3018	1.6435	0.0038
11	0.9754	0.3357	1.6150	0.0036
12	0.9776	0.3652	1.5899	0.0035
13	0.9794	0.3911	1.5899	0.0033
14	0.9810	0.4140	1.5479	0.0032
15	0.9823	0.4346	1.5300	0.0031
16	0.9835	0.4532	1.5138	0.0030
17	0.9845	0.4700	1.4990	0.0030
18	0.9854	0.4854	1.4854	0.0029
19	0.9862	0.4995	1.4729	0.0028
20	0.9869	0.5126	1.4613	0.0028
21	0.9876	0.5247	1.4505	0.0027
22	0.9882	0.5350	1.4404	0.0027
23	0.9887	0.5464	1.4310	0.0026
24	0.9892	0.5562	1.4222	0.0026
25	0.9896	0.5654	1.4139	0.0025

[a]The probability of $s_{\bar{X}}$ values above the upper control limit for $s_{\bar{X}}$ when sampling from a normal distribution with $s_{\bar{X}} = s_{\bar{X}}/\sqrt{n}$ have been calculated from tables of the χ^2 distribution given in *Handbook of Mathematical Functions*, Milton Abramowitz and Irene A. Stegun, eds., National Bureau of Standards, Applied Mathematics Series, No. 55, 1972, using the interpolation formulas provided with the table.

ADJUSTMENT STRATEGIES

Process adjustments are made to try to keep the process centered on the basic dimension. Why should the process exhibit
shifts from this centering? Gradual shifts may occur because
of such factors as tool wear, nonuniformity of raw material, and
temperature variations. Abrupt shifts may occur because of
changes in material (one bar of stock to the next), dress/dress
compensation mismatches, local variations in tool hardness, etc.
In principle, all of these are assignable causes which could be
identified and removed or, at least, minimized. Some may be
so systematic that they can be reliably predicted and programmed
compensations can be made.

Consideration here is limited to the special case in which

(a)   Prompt adjustment is the best or only available response,
(b)   The need for an adjustment cannot be anticipated, and
(c)   The shifts are obscured by random variation.

If (a) is not true, then process improvements must be made.
If (b) is not true, then programmed adjustments should be made
to control the process. If (c) is not true, then either the
shifts are too small to be detected and control is not possible
or the shifts are so large as to be readily detected and control
is straightforward compensation for these shifts. Thus the
special case is the case of particular interest.

Consider, then, a process for which shifts occur, random
variation is significant, and adjustments are possible to keep
the process centered on the basic dimension. Initially the adjustment is set at zero. The first unit is produced and gauged
to determine the result, X_1. This result is not identical to the
basic dimension, X', because some random variation has occurred,
but the result is close enough that no out-of-adjustment signal
is generated. Therefore, the adjustment is left at zero. The
process continues to run, the ith unit, X_i, generates an out-of-
adjustment signal, and, as a result of this, an adjustment Δ_i
is made. The process continues to run and unit X_{i+1} is produced. This is summarized in Table 10.2.

The point of Table 10.2 is to make it absolutely clear that
the current result, X_{i+1}, is determined by random variation
plus any shifts that have occurred plus the previous results
which have determined the current adjustment. In this case

TABLE 10.2   The Delayed Effect of Out-of-Adjustment Signals

Observation number	Adjustment	Result	Evaluation	Action	New adjustment
1	0	X_1	OK	None	0
2	0	X_2	.	.	.
.	.	.	.	.	.
.	.	.	.	.	.
.	.	.	.	.	.
i	0	X_i	Out	Adjustment	Δ_i
i + 1	Δ_i	X_{i+1}	.	.	.

the introduction of the adjustment strategy has made the results recursive, that is, dependent on the foregoing results.   For actual processes this may be true in any case for reasons other than the adjustment strategy.

When a process is recursive it is less stable, readily driven into oscillation by inappropriate adjustment strategies.   If the process is inherently recursive, proper control is best achieved by designing a digital filter based on the specific characteristics of the process.   Digital filter theory is beyond the scope of this discussion; the reader faced with the problem of an inherently recursive process should consult tests on this subject (Ref. 15).

If the problem is due solely to the adjustment strategy, then it is reasonable to try to estimate the result, Y_{i+1}, which would have been obtained if no adjustments had been made.   Assuming that the adjustments are linear over the range of interest and that the random variation in the adjustment mechanism is small compared to the overall random variation of the process, it is reasonable to calculate

$$Y_{i+1} = X_{i+1} - \Delta_i \tag{10.12}$$

and use these values to calculate new adjustments as they are necessary.   It should be kept clearly in mind that the X's are the actual results and must be used to determine whether the results are out-of-specification, are out-of-adjustment, are out-of-control, and to determine the statistics of the process.   The

TABLE 10.3   The Effect of Adjustment Sample Size

n	$\sqrt{n}$	$P(\sqrt{n})$	$2 * 10^6 * P(\sqrt{n})$
4	2	0.022750	45,500
7	2.646	0.004076	8,151
16	4	0.000032	63
20	4.472	0.000004	0

Y's are introduced only for the purpose of calculating the adjustments to be made.

A simple adjustment strategy is to base the current adjustment on the difference between the previous Y values and the
basic dimension, X', since the Y's are the expected results if
no adjustments were made and the object is to correct for any
shift from the basic dimension. Since the Y's contain significant random variation, it is best to average them to avoid over-
responding to any extreme individual variation. Thus the following formula for the adjustment might be used:

$$\Delta_i = \frac{\Sigma_{j=i-n+1}^{i} (Y_j - X')}{n} \qquad (10.13)$$

where n is the sample size. If the assumptions made are correct, the Y's will have the same standard deviation, s_X, as the
X's and the Δ's will have a standard deviation $s_X/\sqrt{n}$. It is
clear that this adjustment strategy will move a stable process
which is initially centered on the basic dimension off this perfect centering in response to the random variations in the process output. Assuming the process is perfectly centered, it is
easy to calculate the probability that the next adjustment will
shift the process a full standard deviation, s_X, or more from
its proper centering. Such a shift amounts to $\sqrt{n}$ standard
deviations for the adjustment, so, assuming the adjustments are
normally distributed, Table 10.3 summarizes some typical results.

In a running set of adjustments only a few will be made
from the starting point of perfect centering. Thus the last
column in Table 10.3 is only roughly indicative of the number
of times in ppm that the adjustment strategy defined above
might shift a perfectly stable process a full standard deviation

from its proper centering. Nonetheless, from this it is clear
that unnecessary large adjustments will be minimized by selecting as large a value of n as possible.

EVALUATING COMBINED OUT-OF-ADJUSTMENT
TESTS AND ADJUSTMENT ALGORITHMS

To pull all of the foregoing points together, a set of 160 tests
of combined out-of-adjustment tests and adjustment strategies
were conducted using eight sets of test data as follows:

1. Normal Data: 100,000 normally distributed random numbers
                  (mean = 50.02, standard deviation = 4.99)
2. Step1.25:   Normal Data plus steps of 1.25 at intervals
                  of 100 numbers increasing to 6.25, then
                  decreasing to -6.25.
3. Ramp1.25:   Normal Data plus a ramp up to 1.25 at the
                  43rd number, down to -1.25 at the 129th
                  number, then increasing to 0 at the 173rd
                  number. This pattern repeats until the
                  100,000th number.
4. Step2.5:   Normal Data plus steps of 2.5 as in Step1.25.
5. Ramp2.5:   Normal Data plus ramps of 2.5 as in Ramp1.25.
6. StepRamp2.5:   Normal Data plus steps of 2.5 as in Step1.25
                  plus ramps of 2.5 as in Ramp1.25.
7. Step5:   Normal Data plus steps of 5 as in Step1.25.
8. Ramp5:   Normal Data plus ramps of 5 as in Ramp1.25.

Five Test & Adjust programs were written which can scan these
data sets, generate out-of-adjustment and out-of-control signals,
make offset adjustments with the formula given above for any
sample size n, and record the number of out-of-specification
results, the number of out-of-control signals, and the number
of out-of-adjustment signals. Four of these programs implement
the basix $\bar{X}$ and $s_{\bar{X}}$, the comprehensive $\bar{X}$ and $s_{\bar{X}}$, the standard
mask Cusum and $s_{\bar{X}}$, and the semiparabolic mask Cusum and $s_{\bar{X}}$
tests as described above. The fifth program makes an adjustment after each observation regardless of the result.

The results of these tests are given in Table 10.4 as a pair
of numbers, specifically the number of out-of-specification results in ppm (the actual result multiplied by 10 since there are

only 100,000 numbers in a data set) and the number of out-of-
control signals generated by the s test per 100 observations.
The specification limits used for the tests were 30 to 70, which
correspond to a c_{pk} of 1.33 for the Normal Data set. The
choices of mean, standard deviation, specification limits, and
reporting of out-of-specification results in parts per million are
arbitrary but reasonable to put the results in the context of
parts per million quality objectives. The choice of reporting
the out-of-control signals per 100 observations was made to
give some feel for the frequency of out-of-control signals to
be expected from a capable and stable process.

Another aspect of these tests which must be pointed out is
that the tests were made on individual observations, not on
averaged observations. In other words, an X and s_X chart is
being used rather than an $\bar{X}$ and $s_{\bar{X}}$ chart. There is no loss
in generality since, in this instance, the random variation is
known to be normally distributed. If this were not the case,
the earlier remarks emphasizing the value of averaging because
the distribution of $\bar{X}$ is always nearer to the normal distribution
than is the distribution of X would apply. Since there is no
loss of generality in this special case, the choice of the X and
s_X chart was made to maximize the number of observations
tested. The same comments apply equally to the Cusum charts,
though the length of the mask could be shortened if control is
to be based on $\bar{X}$ rather than X. The standard deviation $s_{\bar{X}}$
must, of course, be estimated from a sample and in all cases
a sample size of five was used, consisting of the current and
the four previous observations. The results of the tests may
be summarized as follows.

The Reference Test

All of the Test & Adjust programs can be run as Test Only programs. If this is done on the Normal Data set, the result must
be and is always the same, namely 80 ppm out-of-specification
observations compared to 64 for a perfect normal distribution
and 0.675 out-of-control signals compared to 0.65 for a perfect
normal distribution.

The Base Line Test

If the Test & Adjust programs are run on the Normal Data set
with the adjustment algorithm enabled, the adjustments in

response to random variation result in increased out-of-specification observations and out-of-control signals. If the sample
size for the adjustment calculations is 16 or 20, the number of
out-of-specification observations roughly doubles; for a sample
size of four it is roughly five times the reference test level.
The number of out-of-control signals rises by roughly 50% for
adjustment sample sizes of 16 and 20 and by roughly 300% for
an adjustment sample size of 4.

The Control Test

In Table 10.4 the lowest two values for the number of out-of-
specification observations have been underlined for each data
set. A number of features become apparent:

The 100% adjustment program clearly has the best performance overall. For each adjustment sample size, little or no
degradation in performance is seen until steps and ramps of
the magnitude of a full standard deviation, s_X, are encountered.

The semiparabolic mask Cusum program is usually a close
second to the 100% adjustment program. In fact, they are very
similar in that even on Normal Data the semiparabolic mask Cusum program made 23,000 to 26,000 adjustments, and this only
rises to 29,000 for the worst case in the Step5 data set. In
other words, it is really a 25% adjustment program because of
frequent triggering by the random variation component of the
data sets.

The basic X and s program clearly has the worst performance
overall. It triggers very infrequently, typically on 0.5 to 1.5%
of the observations. Since its trigger mechanism is a single, low-
probability event, it seems reasonable that it would tend toward
undercontrol.

For small steps and ramps an adjustment sample size of 20
is best; for larger steps and ramps an adjustment sample size
of 16 becomes preferable. This is to be expected since, if the
process is shifting rapidly, an adjustment algorithm based in
part on much earlier data is likely to underestimate the magnitude of the shift.

All of the programs except the basic $\bar{X}$ and $s_{\bar{X}}$ program are
able to perform so well that no increase is seen in number of
out-of-control signals over those generated on the Normal Data

TABLE 10.4   Test Results for Five Test and Adjust Programs

n for adj.	Method	No adjustment N(50, 5)	N(50, 5)	Step 1.25
	Adj. 100%	80 / .7	400 / 3.5	400 / 3.5
	Cusum—SP	80 / .7	370 / 3.1	360 / 3.1
4	X & s—Compr.	80 / .7	410 / 2.7	350 / 2.6
	Cusum—Std.	80 / .7	280 / 2.4	420 / 2.6
	X & s—Basic	80 / .7	410 / 1.7	520 / 1.8
	Adj. 100%	80 / .7	190 / 1.7	190 / 1.7
	Cusum—SP	80 / .7	170 / 1.7	160 / 1.7
7	X & s—Compr.	80 / .7	250 / 1.5	250 / 1.5
	Cusum—Std.	80 / .7	230 / 1.4	240 / 1.5
	X & s—Basic	80 / .7	270 / 1.2	350 / 1.3
	Adj. 100%	80 / .7	150 / 1.0	150 / 1.0
	Cusum—SP	80 / .7	160 / 1.0	160 / 1.0
16	X & s—Compr.	80 / .7	160 / 1.0	180 / 1.0
	Cusum—Std.	80 / .7	<u>130</u> / .9	<u>140</u> / .9
	X & s—Basic	80 / .7	<u>130</u> / .9	260 / 1.0
	Adj. 100%	80 / .7	140 / .9	<u>140</u> / .9
	Cusum—SP	80 / .7	140 / 1.0	<u>140</u> / 1.0
20	X & s—Compr.	80 / .7	<u>130</u> / .9	120 / .9
	Cusum—Std.	80 / .7	<u>100</u> / .8	<u>140</u> / .8
	X & s—Basic	80 / .7	170 / .8	350 / 1.0

[a]The first number in each pair is the reported number of out-of-specification results in parts per million. The second number is the reported number of out-of-control in s signals per 100 observations. The two smallest out-of-specification results in each column are underlined. Note that minor differences in these results can occur from column to column and row to row because the out-of-adjustment signals cause shifts at different time points in the data sequences. Attention should be focused on trends and large differences.

Conducted for Four Different Adjustment Sample Sizes[a]

Adjust offset

Ramp 1.25	Step 2.5	Ramp 2.5	StepRamp 2.5	Step 5	Ramp 5
380 / 3.5	400 / 3.5	370 / 3.5	370 / 3.5	440 / 3.6	340 / 3.5
310 / 3.1	360 / 3.1	330 / 3.1	340 / 3.2	430 / 3.2	350 / 3.2
390 / 2.6	360 / 2.7	410 / 2.8	430 / 2.8	470 / 3.0	500 / 2.9
390 / 2.4	420 / 2.8	480 / 2.8	510 / 2.8	510 / 3.2	500 / 2.9
380 / 1.8	610 / 2.0	660 / 2.1	720 / 2.3	950 / 2.5	1190 / 2.7
180 / 1.7	200 / 1.7	170 / 1.7	180 / 1.7	270 / 1.8	180 / 1.7
150 / 1.7	160 / 1.7	150 / 1.7	170 / 1.7	250 / 1.8	180 / 1.7
280 / 1.5	250 / 1.6	300 / 1.5	330 / 1.6	300 / 1.7	290 / 1.6
220 / 1.5	230 / 1.5	260 / 1.5	350 / 1.6	450 / 1.7	360 / 1.6
290 / 1.3	450 / 1.4	420 / 1.5	600 / 1.6	640 / 1.8	1050 / 1.9
140 / 1.0	170 / 1.0	150 / 1.0	160 / 1.0	230 / 1.1	170 / 1.0
160 / 1.0	180 / 1.0	160 / 1.0	210 / 1.0	280 / 1.1	180 / 1.0
150 / 1.0	190 / 1.0	250 / 1.0	220 / 1.0	340 / 1.1	220 / 1.0
160 / .9	160 / .9	180 / .9	240 / .9	200 / 1.0	300 / 1.0
250 / .9	450 / 1.2	420 / 1.0	670 / 1.2	880 / 1.5	950 / 1.5
140 / .9	140 / 1.0	150 / .9	180 / 1.0	250 / 1.0	210 / .9
160 / 1.0	140 / 1.0	180 / 1.0	210 / 1.0	260 / 1.0	210 / 1.0
90 / .8	160 / .9	240 / .9	240 / .9	270 / 1.0	300 / 1.0
150 / .8	180 / .9	230 / .9	300 / .9	300 / .9	240 / .9
250 / .9	410 / 1.1	390 / 1.0	550 / 1.2	850 / 1.4	940 / 1.4

TABLE 10.5   The Increase in $s_{\bar{X}}$ Out-of-Control Signals with Loss of Capability[a]

n for adj.	Method	$s_{\bar{X}} = 5$, N(50, 5)	$s_{\bar{X}} = 4$, N(50, 5)	Ratio of out-of-control signals
	Adj. 100%	400 / 3.5	370 / 15.4	4.4
	Cusum—SP	370 / 3.1	340 / 14.3	4.6
4	X & s—Compr.	410 / 2.7	340 / 12.1	4.5
	Cusum—Std.	280 / 2.4	420 / 12.5	5.2
	X & s—Basic	410 / 1.7	340 / 9.1	5.4
	Adj. 100%	190 / 1.7	180 / 10.0	5.9
	Cusum—SP	170 / 1.7	170 / 9.8	5.8
7	X & s—Compr.	250 / 1.5	230 / 8.7	5.8
	Cusum—Std.	230 / 1.4	180 / 8.8	6.3
	X & s—Basic	270 / 1.2	220 / 7.4	6.2
	Adj. 100%	150 / 1.0	140 / 7.3	7.3
	Cusum—SP	160 / 1.0	150 / 7.3	7.3
16	X & s—Compr.	160 / 1.0	160 / 6.7	6.7
	Cusum—Std.	130 / .9	130 / 6.7	7.4
	X & s—Basic	130 / .9	160 / 6.3	7.0
	Adj. 100%	140 / .9	120 / 6.8	7.6
	Cusum—SP	140 / 1.0	110 / 6.9	6.9
20	X & s—Compr.	130 / .9	150 / 6.5	7.2
	Cusum—Std.	100 / .8	130 / 6.5	8.1
	X & s—Basic	170 / .8	160 / 6.2	7.8

[a]The first column shows the out-of-specification reports in ppm and the $s_{\bar{X}}$ out-of-control reports per 100 observations when the control limits are calculated from the known standard deviation of the process. The second column shows the results when the control limits are calculated from a value which is only 80% of the known standard deviation of the process, thus simulating a loss of process capability. The final column shows the multiple by which the out-of-control signals increased in the example.

set even for the extreme cases of the Step5 and Ramp5 data
sets.

Table 10.5 shows the magnitude of the change in the out-
of-control signals when the upper control limit for $s_{\bar{X}}$ is calculated for $s_{\bar{X}} = 4$ and the programs are run on the Normal Data
set for which $s_{\bar{X}} = 4.99$. With adjustment sample sizes of 16
and 20 it is seen that seven times as many out-of-control signals are generated.

CONCLUSIONS

Based on these results the following recommendations are made:

Automated Control

- Use 100% sampling if the cycle time of the gauge can match
  or exceed the cycle time of the process.
- If at all feasible, use 100% adjustment in response to each
  sample.
- Use an adjustment sample size of the order of 16. If the
  gauge cycle time is significantly slower than the process
  cycle time, use the standard mask Cusum program to test
  the sample result to minimize the number of out-of-adjustment signals without seriously degrading the prompt detection of process shifts. When an out-of-adjustment signal is
  received, gauge extra pieces if necessary to provide the
  data to calculate an optimum adjustment.
- Use the s chart and switch to continuous gauging in response
  to out-of-control signals to guarantee quality. A real degradation of process capability should cause a significant increase in the frequency of out-of-control signals which will
  quickly be detected and should be cause to stop production
  until the cause is identified and corrected.

Do these results imply that $\bar{X}$ and R chart do not work?
Absolutely not! The effort here has been to push the method
to the limit to obtain the best parts per million quality levels
possible. Traditional $\bar{X}$ and R manual charting was initially
applied to processes producing 1% (10,000 ppm) or more out-of-
specification product. The initial effort was to improve the process capability to a c_{pk} of at least 1.33 and, when this was
achieved, to control the process to keep the out-of-specification
product below 0.27% (2700 ppm). Even the worst case in Table

10.4, a basic $\bar{X}$ and $s_{\bar{X}}$ test and adjustment sample size of 4, does better than that! (The manual method is more likely to calculate the adjustments from an average of actual results which could be recursive if several samples are averaged and adjustments occurred between samples. This would likely give less satisfactory results and should be avoided.)

If control must be by manual charting the following recommendations are made.

Manual Control

• Use the comprehensive $\bar{X}$ or standard mask Cusum test. These generate out-of-adjustment signals approximately 4 and 2% of the time, respectively, from the Normal Data set and approximately 5.5 and 4.5% of the time, respectively, from Step2.5 and Ramp2.5 data. The results are not as good as would be obtained with 100% adjustment but, for any given sample size, they hold up reasonably well out to Step2.5 and Ramp2.5. If shifts and drifts larger than these are being encountered, process improvement is clearly required, particularly if c_{pk} is marginal.

• Do Not Rely on the Basic $\bar{X}$ test!

• Use the conventional R chart rather than the $s_{\bar{X}}$ chart. Out-of-control signals of the order of 1 to 2 per 100 observations are to be expected from random variations but should not be ignored. A possible strategy is to switch to continuous or very frequent sampling to guarantee product quality while evaluating the situation. A real degradation of process capability should cause a significant increase in the frequency of out-of-control signals, which will quickly be detected and should be cause to stop production until the cause is identified and corrected.

In summary, these tests emphasize the fundamental limitations of process control when parts per million quality levels are essential. Even a reasonably stable process with a c_{pk} of 1.33 is marginal for this requirement and is a good candidate for further process improvement. If the process is inherently recursive, even an adjustment strategy based on a well-designed digital filter can be expected to exhibit more amplification of the random variation than is seen in the idealized test cases of Table

10.4. Still, excellent parts per million quality levels are attainable and early detection and correction of small process shifts obscured by the noise of random variation is possible.

REFERENCES

1. The Strategic Planning Institute, 955 Massachusetts Ave., Cambridge, MA 02139, Tel. (617) 491-9200, has conducted extensive research on this subject.
2. R. J. Schonberger, "World Class Manufacturing," Free Press, New York, 1986.
3. R. Hall, "Attaining Manufacturing Excellence, Dow Jones Irwin, Homewood, Illinois, 1987.
4. The Association for Manufacturing Excellence, PO Box 504, Elm Grove, WI 53122, Tel. (414) 785-0970, is an excellent source of current information.
5. R. Hall, "Just-in-Time-Production in den USA," in Just-in-Time Produktion, Prof. Dr. Horst Wildemann, Handelsblatt Gmbh, Postfach 11 02, Kasernenstr. 67, 4000 Duesseldorf 1, Germany.
6. G. Taguchi, "Introduction to Quality Engineering," Asian Productivity Organization, 1986. (Available from the American Supplier Institute, PO Box 42424, Detroit, MI 48242, Tel. (313) 728-4110.)
7. See also articles on the Taguchi methods in the Journal of Quality Technology, Vol. 17, No. 4, October 1985, and in Quality Progress, Vol. XX, No. 6, June 1987.
8. N. L. Johnson and F. C. Leone, "Statistics and Experimental Design in Engineering and the Physical Sciences," 2nd ed., Vol. I, p. 153, Wiley, New York, 1977.
9. ANSI Z1.1-1958(R1975) and Z1.2-1958(R1975), American National Standards Institute, 1430 Broadway, New York, NY 10018. See paragraph 4.1, p. 11: "A control chart for R may be used as a substitute for a control chart for sigma with but little loss in efficiency...." With current microprocessors there is no reason to sacrifice efficiency. (Note: statisticians generally use the letter s for the standard deviation of the sample and reserve the Greek letter σ for the population parameter. Earlier quality control texts use σ for the sample statistic.)
10. See any standard reference on statistical quality control, e.g., "Statistical Qualify Control," Eugene L. Grant and Richard S. Leavenworth, McGraw-Hill, New York, 1980;

"Statistical Quality Control Handbook," Western Electric
(available through AT&T Customer Information Center, P.O.
Box 19901, Indianapolis, IN); "Total Quality Control,"
Armand V. Feigenbaum, McGraw-Hill, New York, 1983;
"Quality Planning and Analysis," J. M. Juran and Frank
M. Gryna, Jr., McGraw-Hill, New York, 1980.

11.   This discussion in terms of analysis of variance—one factor and k levels (one-way classification)—will be found in
basic texts on the subject. The treatment of the subject
in "Statistics and Experimental Design in Engineering and
the Physical Sciences," Vol. II, Norman L. Johnson and
Fred C. Leone, Wiley, New York, 1977, is particularly
good. The relevant material is to be found on pp. 593-
611, the opening pages of the volume. The reader should
note that this reference uses the uppercase letter S for
the sum of the squares, while the notation herein uses the
lowercase letter s for the standard deviation.

The design, process, or quality engineer who is not
conversant with the analysis of variance should study this
topic as a powerful tool for the design and analysis of efficient experiments to determine optimum values for basic
dimensions, to determine appropriate tolerances for these
dimensions, and to improve process capabilities where this
is essential to hold functional specifications. For this purpose the text "Statistics for Experimenters," George E. P.
Box, William G. Hunter, and J. Stuart Hunter, Wiley, New
York, 1978, is a particularly recommended as eminently
readable with a strong, clear focus on the objectives of
the procedures.

12, $vs_{\bar{x}}^2/\sigma^2$ has a χ^2 ψ distribution where v is the number of
degrees of freedom and σ^2 is the variance of the population.
See W. E. Deming, "Some Theory of Sampling," Wiley, New
York, 1950, p. 150ff for a discussion with historical references of the distribution of the standard deviation of the
normal distribution.

13.   See Grant and Leavenworth, p. 85, for a discussion of
this approximation.

14.   See Johnson and Leone, p. 348.

15.   See, for example, "Time Series Analysis: Forecasting and
Control," George E. P. Box and Gwilym M. Jenkins, Holden
Day, 1976, chapters 12 and 13. "Digital Filters," R. W.
Hamming, Prentice-Hall, Englewood Cliffs, N.J., 1977,

provides an excellent introduction to digital filters and chapter 3, in particular, reexamines the classical polynomial approach to demonstrate the added insight provided by the frequency approach on which digital filter theory is based.

Index